Party

From its inception in 2001, the United Russia party has rapidly developed into a hugely successful, organisationally complex political party and key component of power. This book provides a much needed analysis on United Russia by exploring the role of the party in the Russian political system, from 2000 to 2010. It explores the party empirically, as an impressive organisation in its own right, but also theoretically, as an independent or explanatory variable able to illuminate the larger development of dominant-power politics in Russia in the same period.

The book creates a model to understand the role of political parties in electorally based political systems and shows how United Russia conforms to this model, and, importantly, how the party also has unique features that affect its place in the political system. The book goes on to argue that, as an outcome of political changes occurring elsewhere, the United Russia party represents a reversal of the typical relationship between parties and power found in comparative literature. This has potentially far-reaching implications for our understanding of party dominance in the twenty-first century and also the sources of regime stability and instability.

ean P. Roberts is a Visiting Researcher at the Norwegian Institute of International Affairs (NUPI) in Oslo.

BASEES/Routledge Series on Russian and East European Studies

Series editor:
Richard Sakwa, Department of Politics and International Relations, University of Kent

Editorial Committee:
Julian Cooper, Centre for Russian and East European Studies, University of Birmingham

Terry Cox, Department of Central and East European Studies, University of Glasgow

Rosalind Marsh, Department of European Studies and Modern Languages, University of Bath

David Moon, Department of History, University of Durham

Hilary Pilkington, Department of Sociology, University of Warwick

Graham Timmins, Department of Politics, University of Stirling

Stephen White, Department of Politics, University of Glasgow

Founding Editorial Committee Member:
George Blazyca, Centre for Contemporary European Studies, University of Paisley

This series is published on behalf of BASEES (the British Association for Slavonic and East European Studies). The series comprises original, high-quality, research-level work by both new and established scholars on all aspects of Russian, Soviet, post-Soviet and East European Studies in humanities and social science subjects.

1. Ukraine's Foreign and Security Policy, 1991–2000
Roman Wolczuk

2. Political Parties in the Russian Regions
Derek S. Hutcheson

3. Local Communities and Post-Communist Transformation
Edited by Simon Smith

4. Repression and Resistance in Communist Europe
J.C. Sharman

5. Political Elites and the New Russia
Anton Steen

6. Dostoevsky and the Idea of Russianness
Sarah Hudspith

Putin's United Russia Party

Sean P. Roberts

Routledge
Taylor & Francis Group

LONDON AND NEW YORK

First published 2012
by Routledge
711 Third Avenue, New York, NY 10017

Simultaneously published in the UK
by Routledge
2 Park Square, Milton Park, Abingdon, Oxon OX14 4RN

Routledge is an imprint of the Taylor & Francis Group, an informa business

First issued in paperback 2013

British Library Cataloguing in Publication Data
A catalogue record for this book is available from the British Library

Library of Congress Cataloging in Publication Data
Roberts, S. P.
Putin's United Russia Party / S. P. Roberts.
p. cm. — (BASEES/Routledge series on Russian and East European studies
; 77)
Includes bibliographical references and index.
1. Politicheskaia partiia "Edinaia Rossiia". 2. Political parties—Russia
(Federation) I. Title.
JN6699.A8P6545 2011
324.247'04–dc23
2011020059

ISBN: 978–0–415–66902–3 (hbk)
ISBN: 978–0–415–72830–0 (pbk)
ISBN: 978–0–203–18091–4 (ebk)

Typeset in Times New Roman by Prepress Pojects Ltd, Perth, UK

Contents

Illustrations

Figures

Tables

Acknowledgements

Very few of our achievements are really our own, and so to acknowledge this fact I would like to thank the following for their professional support. They include a special thanks to David White (CREES, University of Birmingham) for his patient supervision and for sharing his ideas on Russian party politics and on Kidderminster Harriers. Sincere thanks also to Derek Averre, Marea Arries, Tricia Carr, Veta Douglas (CREES) and all the staff at CREES. I would like to show my appreciation to the following researchers from the Russian and Eurasian Department at the Norwegian Institute of International Affairs (NUPI) for their help. They are Konstantin Anchin, Helge Blakkisrud, Geir Flikke, Jakub Godzimirksi, Victor Jensen, Nino Kemoklidze, Indra Øverland, Natalia Moën, Julie Wilhelmsen and Elena Wilson Rowe. I extend my thanks to everyone at NUPI for their friendly disposition during my stay in Oslo. Elsewhere, thanks to Karen Henderson, Mark Phythian, Renie Lewis and Zoe Clapton (University of Leicester) for immediate post-PhD support and Anne White, Howard White and Rosalind Marsh (University of Bath) for pre-PhD motivation. I would also like to acknowledge the assistance of the UK Economic and Social Research Council (ESRC) and the Norwegian Research Council for my overall academic development. I extend my special thanks to every expert and political figure in Moscow who gave up their time to speak to me during fieldwork, in particular Andrei Loginov and Vladimir Pribylovskii whose detailed knowledge of Russian politics was deeply impressive. Naturally, all faults and shortcomings with this work are entirely my own.

I must finish by thanking a number of special people behind the scenes for their considerable help in enabling me to start, persevere with and finish this project. Thanks to Rich, Rob and Alexandra for the fireside chats, as too, Thomas 'Steve' Shepherd, who passed away in 2006, but who is not forgotten. Massive thanks to Mila and Galina Petrovna for all their help in Moscow and for their continuing friendship. Finally, thanks to Ulrike, mum, dad, Sarah, George and Mollie for all the support you gave along the way and for helping to keep things in perspective.

Sean Roberts

Abbreviations

CEC	Central Election Commission of the Russian Federation
CEJ	Centre for Extreme Journalism
CPD	Congress of People's Deputies
CPRF	Communist Party of the Russian Federation
CPSU	Communist Party of the Soviet Union
EMERCOM	Ministry for Emergency Situations
EPP	European People's Party
FAR	Fatherland All-Russia
FOM	Public Opinion Foundation
FSB	Federal Security Service
KGB	Committee for State Security
KMT	Kuomintang
LDP	Liberal Democratic Party (Japan)
LDPR	Liberal Democratic Party of Russia
NATO	North Atlantic Treaty Organization
ODIHR	Office for Democratic Institutions and Human Rights
OSCE	Organization for Security and Co-operation in Europe
PRES	Party of Russian Unity and Accord
PRI	Partido Revolucionario Institucional
RAO UES	Russian Joint-Stock Company, Unified Energy System
RSFSR	Russian Socialist Federative Soviet Republic
SMD	Single-Mandate Districts
UMNO	United Malays National Organisation
USSR	Union of Soviet Socialist Republics
VTsIOM	Russian Public Opinion Research Centre

Preface

This case study examines the role of United Russia in the rise of dominant-power politics, also termed electoral authoritarianism, which characterises the post-Yeltsin period (2000–10). Comparative literature identifies parties as important explanatory variables in a range of regime outcomes, including the successful consolidation of democracy, but also the consolidation and persistence of authoritarian rule. The impressive rise of United Russia in the Russian political system from December 2001 onwards, together with its co-occurrence with the growing strength of the post-Yeltsin regime, suggests that the party was a key factor in the outcome of the latter.

This research first develops a theoretical framework to understand the role of ruling parties in modern political systems and then applies this framework to explore the Russian case. Although a component of power in the Putin and Medvedev periods, collectively termed the post-Yeltsin period, this book argues that the origins of United Russia in the party of power phenomenon limit its value as an explanatory variable. Rather than a principal power in the emerging post-Yeltsin political order, United Russia is an agent of a powerful civilian executive, which remains beyond the control of any party. In this sense, the rise of United Russia is misleading. Rather than a source of dominant-power politics, as is the case with many twentieth-century ruling parties, United Russia is simply a reflection of the intentions and ability of non-party power-holders to project their power onto party agents. This research contributes to existing literature on party politics in Russia and the post-Soviet space and, in comparative terms, to the dynamics of party and regime dominance.

Introduction

In 2002, an article by Thomas Carothers appeared in the *Journal of Democracy*, identifying two major tendencies within the third wave of democratic transitions. As most transition countries, Carothers observed, are neither dictatorial nor heading for democracy, 'feckless pluralism' describes those states where democracy remains 'shallow and troubled'; while 'dominant-power politics' relates to states where, despite the presence of democratic institutions, one political group dominates power, undermining any sense of power contestation and alternation (Carothers 2002: 9–14). This reality, seen in many democratic transitions in the third wave, has, in part, provided the impetus for new studies of authoritarianism in the twenty-first century in nearly every corner of the world (Angrist 2006; Brownlee 2007; Bunce *et al.* 2009; Ezrow and Frantz 2011; King 2010; Levitsky 2003; Levitsky and Way 2010; McCoy and Myers 2006; Ottaway 2003; Rodan 2004; Schedler 2002, 2006a; Way 2006; Zakaria 2004).

In Russia, the Yeltsin period (1991–9) personified feckless pluralism, simultaneously revealing the weakness of the state and Russia's democratic project in general. In contrast, the dominant-power politics of the post-Yeltsin period (2000–10) seems to mark a transformation in the fortunes of Russia's ruling elite. The nature of this transformation remains largely unexplained, although one important regime variable absent in the Yeltsin period but present in the post-Yeltsin period may hold the key – strong, pro-regime party support. In fact, the rise of dominant-power politics in the post-Yeltsin period coincides with the rise of the all-national political party United Russia as a major player in Russian politics.

This study examines United Russia and its contribution to the post-Yeltsin regime in the second decade of post-Soviet politics. The aim is to clarify political processes in this period and highlight possible pathways of future political development, but also to refine our understanding of ruling parties and their relationship with larger, regime outcomes. The following paragraphs outline these aims as well as the underlying rationale.

Aims and rationale

This book is essentially a study of ruling parties and their value as independent or explanatory variables, but also an exploration of dominant-power politics in Russia, as the dependent variable of interest. As such, the following chapters address two main questions pertaining to (1) United Russia's exact relationship with dominant-power politics and (2) the party's overall role in the post-Yeltsin period. As noted, the focus is on the second decade of post-Soviet politics (2000–10), also termed the 'post-Yeltsin period', which in itself signifies the strength of political change occurring after Yeltsin stepped down as president of the Russian Federation on New Year's Eve, 1999, but also the difficulty of placing the 'Putin period' within any concrete time frame. Although the Putin period officially encompasses his two presidential terms (2000–8), it is unclear that his influence over politics has actually abated since the election of Dmitrii Medvedev as president of the Russian Federation in March 2008, making it difficult to identify a clear cut-off point for the Putin period. The terms 'regime' and 'political order' are used interchangeably to refer to the groups and individuals who hold power in the political system – the nature of which becomes clear as this study progresses.

The rationale for this study can be summarised in terms of the puzzle of United Russia. This puzzle, put simply, is that the rise of this party as a dominant force in Russian politics, together with the substantial comparative literature that identifies parties as key variables in regime outcomes, strongly suggests that the party is central to the political transformation occurring between the Yeltsin and post-Yeltsin periods. However, the background of weak party development in the Yeltsin period and the larger backdrop of de-partification following the collapse of the Soviet Union raises an important secondary question – how was the development of an electorally successful, even dominant party possible in view of this recent Russian party political history? The absence of detailed research on this subject leaves these questions largely unresolved. The following paragraphs outline these points in more detail.

The rise of United Russia

The rise of United Russia can be traced to the equally sharp rise of its predecessor, Unity (Edinstvo), from September 1999, at the very end of the Yeltsin period. Unity achieved significant and largely unexpected success in the election to the lower chamber of the national legislature (State Duma) in December 1999, providing a platform for the successful election of Vladimir Putin as president in March 2000. Unity then merged with its main competitor, the Fatherland All-Russia bloc (Otechestvo Vsya-Rossiya – FAR), in December 2001 to form the All-Russian Party of Unity and Fatherland – United Russia (Vserossiiskaya Politicheskaya Partiya Edinstvo i Otechestvo – Edinaya Rossiya), later renamed United Russia (Edinaya Rossiya) following the party's IV Congress in December 2003. United Russia's subsequent development has been remarkable. Within a relatively short period, the party achieved impressive electoral success in both national and regional parliamentary elections, squeezing out opposition forces and providing stable majorities for the emerging regime. In the State Duma election of 2003, arguably the first confirmation of dominant-power politics, the party collected 49.3 per cent of the popular vote, gaining 222 out of a possible 450 parliamentary seats. For the State Duma election of December 2007, the party gained 64 per cent of the popular vote, giving the party a constitutional majority of 315 seats (Russia Votes n.d.).

However, the rise of United Russia is by no means confined to majorities in the State Duma or success in legislative elections. For some time now, articles appearing in Russian and international media have noted growing similarities between the post-Yeltsin ruling party and the ruling party of the Soviet period in the form of the Communist Party of the Soviet Union (CPSU), creating a perception, at least, that the post-Yeltsin political order has turned back the clock, using the 'Party' as a means to institutionalise rule and generate stability.

An important element of this 'sovietisation thesis' is found in United Russia's overall party development. From early 2002, United Russia rapidly acquired organisational complexity, extensive territorial penetration and relatively high levels of party membership, drawing direct comparisons with the CPSU (Kagarlitsky 2003, 2006; Olenich 2006; Smith 2006; Weir 2006). By 2010, and according to the most recent party figures available, United Russia has party branches in every subject of the Russian Federation, boasting 2,597 district offices and 53,740 local offices (Edinaya Rossiya 2007a). Membership alone makes the party stand out from the general experience of party development in the post-Soviet period. By the presidential election of March 2008, party membership reached two million, equating to an average increase of over 300,000 members per year, every year, for the party's first six years of existence.

Like the CPSU in the Soviet period, it appears that United Russia also performs certain extra party-functions, or those that are not confined to legislating and electioneering. In 2006, the head of the *Perovo* branch of United Russia gave an interview in which he stated that electoral work represented only a small part of the party agenda, an agenda that includes 'work with the population, helping pensioners, orphans, and other vulnerable groups', among other tasks (Weir 2006: 3). In addition, United Russia's successful elite recruitment strategy, seen in the way that 65 of 83 regional executives ran on the party ticket in the 2007 State Duma election, suggests that the party is now the primary mechanism for elite circulation and career advancement within the political system.

At face value, a number of developments within the elite appear to confirm the rise of United Russia as a dominant force in Russian politics. On 15 April 2008, Putin's presidential successor, Dmitrii Medvedev, addressed United Russia at its IX party Congress as the 'ruling party' (Medvedev 2008a). The next day, Vladimir Putin was confirmed as party leader, formalising the link between the powerful former president and the party that had previously supported him.

By the end of 2010, inference from a range of circumstantial evidence leads to a conclusion that Russia has experienced some kind of ruling party revival from its twilight at the end of the Soviet period. But the sovietisation thesis does not end there. From May 2008 and with Prime Minister Putin now ensconced as United Russia's leader, some commentators began to note similarities not only between United Russia and the CPSU, but also between the emerging post-Yeltsin order and the Soviet order of the USSR per se. Senior correspondent with Radio Free Europe/Radio Liberty (RFE/RL), Brian Whitmore, captures this mood well with the following observation:

> Analysts say the emerging arrangement is beginning to resemble the Soviet system, in which actual power resided with the Communist Party, and high state posts – like the President, Prime Minister, or leader of parliament – were largely ceremonial. The general secretary of the Communist Party was the country's true ruler.

(Whitmore 2008)

The existence of political and economic pluralism in Russia, along with domestic and international opinion, means that the single-party or totalitarian model is unlikely to be reproduced in the post-Soviet Russia. However, United Russia and the emerging post-Yeltsin order do share many of the characteristics associated with so-called 'hegemonic party autocracies'. A hegemonic party autocracy is a regime dominated by a ruling party, but in which opposition parties are tolerated and permitted to compete in elections, albeit under conditions that strongly favour the ruling party (Magaloni 2006; Sartori 1976: 204). Although power lies with the ruling party, there is nonetheless a discretional allocation of some power to 'subordinate political groups' (Sartori 1976: 205). Rather than the sovietisation thesis, Russia as a hegemonic party autocracy corresponds with the so-called 'Mexicanisation thesis', one that highlights the growing similarities between United Russia and the Partido Revolucionario Institucional (PRI) – the ruling party that dominated Mexican politics for most of the twentieth century (Gel'man 2005, 2006, 2008; Gvosdev 2002; Reuter and Remington 2009: 504; Riggs and Schraeder 2005).

In short, indicators of United Russia's rise in the post-Yeltsin period, continuing analogies with the CPSU and with ruling parties elsewhere, suggest that the party is now a major political force. These indicators provide cursory evidence which supports the assertion that this party is indeed a major explanatory variable in the transformation occurring between the Yeltsin and post-Yeltsin periods, and the perpetuation of the current regime up to 2010 and beyond.

Comparative support: parties as explanatory variables in macro-political outcomes

The nature of the post-Yeltsin order is examined in more detail in the chapters that follow, but at this juncture it is important to make the following observation: the rise of United Russia as a political force coincides with the rise of dominant-power politics in the post-Yeltsin period. In itself, this co-occurrence strongly suggests a causal role for the former in the outcome of the latter, especially when viewed from the perspective of existing literature in the subfields of party politics and regime studies. In general, comparative literature has long afforded a prominent place to political parties as explanatory variables, that is, independent variables affecting a range of macro-level outcomes. Nowhere is this more evident than in the frequently cited relationship between parties and democratic outcomes, in which the former are seen as necessary, if not sufficient, for democracy to occur (Epstein 1980: 8; Schattschneider 1942: 1).

In 2003, Steven Levitsky presented a paper at Columbia University, arguing that more research needs to be conducted on the problems of building and sustaining contemporary authoritarian regimes (Levitsky 2003). However, existing comparative research already strongly points to the importance of ruling parties. Huntington (1968) was one of the first to identify the importance of ruling parties for institutionalising rule and stabilising authoritarian regimes. For Huntington, the no-party state is one 'without the institutional means of generating sustained change and of absorbing the impact of such change' (Huntington 1968: 404). This conclusion finds resonance in more recent studies, such as Geddes (2003), who demonstrates how authoritarian regime longevity is affected by the presence (and absence) of a strong ruling party. According to Geddes, regimes in which power is vested in a ruling party are often more resilient than both personalist and military regimes or regimes in which parties are weak or absent (Geddes 2003: 82).[1] Brownlee's (2007) comparative study of four authoritarian regimes in the second half of the twentieth century shows how strong ruling parties in Egypt and Malaysia were the decisive variable in regime persistence; while in both Iran and the Philippines the weakness of parties was a factor in regime collapse.

Other studies also lend support to the importance of ruling parties as explanations for macro or state level outcomes, this time accounting for divergent regime trajectories in the post-Cold War period. Some of these studies identify ruling parties as key components of incumbent capacity – a major factor in the generation of dominant-power politics (Alexander 2004; Way 2005, 2006). Way even comments directly on the Russian case, noting that, by the end of the Yeltsin period and the beginning of the Putin period, 'increased organisational capacity' and the presence of a 'disciplined ruling party' led directly to greater 'regime closure' (Way 2005: 248). By 2010, scholars were already examining United Russia as a test case suitable for building a comparative theory of dominant-party formation in an attempt to elucidate the dynamics of authoritarian regime consolidation (Reuter 2010; Reuter and Remington 2009).

In fact, the explanatory power of parties is so great that they have even been identified as decisive factors in macro-economic as well as macro-political outcomes. Corrales (2002) in a study of market reform in Argentina and Venezuela in the 1990s presents a strong argument that political parties represent the key variable in successful economic reform in Argentina, but also reform failure in Venezuela. In Argentina, a cooperative relationship between the ruling party and the executive branch created what Corrales terms the 'state-with-party' condition, which in turn neutralised societal resistance and built credibility for the reformers (Corrales 2002: 32).

Waldner (1999) also acknowledges the importance of ruling parties in explaining longer-term economic outcomes. According to Waldner, relatively low levels of elite conflict in South Korea and Taiwan enabled power-holders in both countries to concentrate on economic development under conditions of political stability, resulting in their spectacular growth from the mid-1960s into the 1980s. However, in Syria (1963–80) and Turkey (1950–80), higher levels of elite conflict necessitated the creation of cross-class coalitions that were later to restrict economic development, as power-holders subordinated long-term economic development to short-term political stability. The key is elite homogeneity, and this is maximised when 'core members of the political elite owe their status to organisational affiliation', and this first and foremost means membership of a strong ruling party (Waldner 1999: 147).

To repeat the comparative logic: the rise of dominant-power politics and the rise of United Russia are likely to be related, with comparative evidence suggesting the latter to be a cause of the former. This logic also gains support when we consider the fact that the feckless pluralism of the Yeltsin period, of shallow and troubled democracy, was accompanied by an absence of a strong ruling party. In sum, there is already a persuasive argument that United Russia is central to the appearance and maintenance or generation of dominant-power politics in the post-Yeltsin period.

United Russia and dominant-power politics: the counter argument

There is no question that the role of United Russia and its place in the political system tell us a great deal about political processes unfolding in the Russian Federation, 2000–10, and that this subject is of interest to Russia-watchers and policy-makers alike. However, this study also provides important clues to Russia's future development. Again, at face value, the presence of United Russia suggests a party-based resolution to the problem of collective action, a problem that plagued the Yeltsin administration, when executive–legislative relations, centre–periphery relations and general elite fragmentation often made governance strained and ineffectual, at times raising doubts over the viability of the Russian state as a whole. In line with the comparative experience already mentioned, if the party has become the means for institutionalising power in the Russian Federation there are strong grounds to expect that dominant-power politics in Russia will persist well beyond the second decade of post-Soviet politics.

There are, however, reasons to doubt the validity of United Russia's electoral and organisational dominance as accurate indicators of its actual relationship with dominant-power politics. The rise of United Russia and its development is somewhat puzzling when viewed from the perspective of overall party development in post-Soviet Russia, especially in the Yeltsin period. There are also the institutional constraints of Russia's 'super-presidential' system to consider (Colton 1995) and the logic of party dominance in a presidential republic. United Russia, as detailed below, appears to be at odds with a great deal of literature from the 1990s that suggests a rather insignificant role for Russian parties.

In the 1990s, a constellation of factors contributed to the overall weakness of parties in Russia, as in many post-Soviet and post-communist states, which in turn reduced their role and importance.[2] A common theme in the literature from this period was the way that the legacies of Soviet rule served to limit the footholds that parties were able to gain in society, conditioning the relationship that parties had with the electorate (Kopecky 1995; Enyedi 2006). Although the impact of communism on party and democracy-sustaining cleavages is difficult to ascertain with any degree of certainty (Whitefield 2002), the weakness of other potential bases of party support, such as civil society, strengthened the more pessimistic assessments of the prospects for strong party development in the post-Soviet space (Ishiyama and Kennedy 2001: 1178).

The absence of durable social bases of party support was mirrored by a general reluctance among many citizens to join parties as members (Toole 2003: 113). This was considered one of the legacies of Soviet and one-party rule, in which the word 'party' evoked negative connotations for many Russian citizens (Bacon 1998: 206). Voters, it was noted, were tired from decades of party appeals to create a new society, both in the former Soviet system and in the newly emerging democratic system (Fleron and Ahl 1998: 241), and these largely exogenous factors, together with the prevalence of modern media technologies as alternative forms of linkage (Lewis 2000: 113), compounded counter-incentives for party-building.

These themes certainly find resonance in the general experience of Russian parties in the 1990s, but also in the specific experience of 'parties of power' in the Yeltsin period. Parties of power, as detailed in Chapter 2, are organisations created from the top down to support executive power. Parties of power in Yeltsin's two terms of office, like most parties, struggled to penetrate the regions (Golosov 2004; Hale 2006; Smyth 2006), failed to attract significant numbers of supporters and achieved only limited success in mobilising voters during elections. On the party-list portion of the 1993 State Duma election, Russia's Choice, the party of the then deputy prime minister, Egor Gaidar, collected just over 14 per cent of the vote, while in 1995, Our Home is Russia, the party of the Prime Minster, Viktor Chernomyrdin, collected only 10 per cent of the vote.

As such, the puzzle of United Russia relates as much to the emergence of dominant-party politics in a post-Soviet context considered by many unsuitable for such a development to occur. This secondary question, of how to account for United Russia's rise, is answered fully in the chapters that follow and forms part of the substantive argument presented in this study. To understand the rise of United Russia it is first necessary to understand the party's exact relationship with dominant-power politics in the post-Yeltsin period. This argument, which is discussed in the next chapter and developed thereafter, relates first and foremost to the interaction between ruling parties, in this case United Russia, and the larger political, historical and social context in which they operate.

Approach and method

This study is a detailed analysis of a single case, one that engages strongly with context, developing theory to guide rather than predict and striving for depth of understanding. As such, the overall approach combines an intensive research design, a significant engagement with history, an explicit theoretical framework and a strong empirical component. Each aspect of this approach, including the rationale, is discussed in the paragraphs that follow.

An in-depth study of United Russia

As shown in the next chapter, comparative studies of ruling parties have furnished valuable insights on regime stability and the nature of power, but these insights are not without problems, especially when context is considered homogeneous. However, there are occasions when comparative research is less suitable, both in terms of the case in focus and the research questions posited. From a practical point of view, across-case comparative research on United Russia is hindered by the lack of existing research on this party and on so-called parties of power in general, while a purely comparative approach would not necessarily shed light on the role of United Russia or its relationship with dominant-power politics. In the final analysis, it is clear that this research is focused on the Russian case first and foremost, and that case studies such as this one are useful in the absence of existing empirical research when little is known of the phenomenon in question (Eisenhardt 1989a: 548).

An in-depth study of a single case is further justified when the case in question has some kind of intrinsic value, and this is certainly true for United Russia on several counts. The first point to note is that United Russia can potentially offer insights on the structure of governance and political development in a country that has a relatively high importance on the international stage. The back cover to Scott Mainwaring's (1999) *Rethinking Party Systems in the Third Wave of Democratization: The Case of Brazil* states that, among the many countries that underwent a transition to democracy in recent times, only Russia is as important to the United States and the larger world as Brazil. Although this study is not primarily aimed at an American audience, the size of Russia, its resources, recent history and impact on the global economy and security go some way to justifying this claim.

The second point to make is that, just as the Russian Federation remains one of the more important countries for the international community, United Russia itself has some intrinsic value as an 'instrumental case' (Creswell and Maietta 2002: 162), that is, a case likely to furnish valuable insights on this type of party elsewhere. As revealed in the chapters that follow, this type of party, whether a party of power or a ruling, dominant or hegemonic party, is one that has a very particular relationship with power. In a similar way, and in terms of dominant-power politics in the post-Cold War period, United Russia presents itself as a 'crucial case', with a potentially high value as an independent or explanatory variable

(George and Bennett 2005: 251), able to shed light on some of the processes of building and maintaining authoritarian regimes. This last point alone justifies its selection for detailed research (Ragin 1987: 22). Finally, United Russia may be said to have intrinsic value as a 'deviant case' (Lijphart 1971) – as a party that seemingly contradicts earlier party development in a specific Russian, and general post-Soviet, context considered unsuitable for the emergence of electorally and organisationally dominant parties. Expressed in different terms, United Russia is a deviant case because of its apparent popularity, seen in its electoral success and relatively high levels of party membership; features that, with the exception of the Communist Party of the Russian Federation (CPRF), run counter to the experience of parties of power in the Yeltsin period.

The drawbacks of single case studies are directly related to their strengths. Although providing detailed, in-depth analysis, they do not contribute to generalisable knowledge and theory development in the same way as comparative research designs (Bennett and George 2001: 137). In response to this shortcoming, this research develops an explicit theoretical position, which, although not comparative across several cases, does use within-case comparison to facilitate understanding and generalisation. This is achieved by the inclusion of a strong historical perspective that creates a distinction between Yeltsin and post-Yeltsin party of power development. In addition, the theoretical framework presented in the next chapter is framed in distinctly comparative terms, while the concluding chapter assesses the comparative implications of this research as a whole. Overall, this case study is concerned with depth of analysis, similar to Weber's call for specialisation (Weber 1946: 137), but grounded in comparative terms, part of an overall conception of area studies to which this study subscribes.

History matters

Historical approaches that ascribe importance to legacy and context are not uncommon in explaining developments in post-Soviet Russia (Bacon 1998; Geddes 1995; Lynch 2005; Roberts and Sherlock 1999; Rose and Shin 2001) and this research is no different. The concern with history reflects its general importance in political analysis, best seen in the so-called state-based and historical institutionalist approaches. State-based approaches focus on larger macro variables, including historical developmental variables, to explain the object of analysis (Shefter 1994; Skocpol 1979). This research on United Russia is also state-based in the sense that a large part of the explanation for the appearance, development and role of United Russia and its relationship with dominant-power politics relates to the way that the Russian Federation emerged from the Soviet Union in December 1991.

The state-based approach is closely aligned with historical institutionalism, which is also relevant for this research. Historical institutionalism suggests that historical choices institutionalise commitments and constrain subsequent action and actor choice. The implication is that if we do not understand initial decisions then it becomes difficult to understand the logic of subsequent social realities

(Guy Peters 2000: 19). A notable feature of this approach is the existence of, and emphasis on, path dependency as a mechanism through which time becomes important. Path dependency, when applied to institutions, shows that, once launched, institutions tend to perpetuate themselves and foreclose other paths of development, even more functionally appropriate paths, unless a sufficiently strong force is able to change this path (Krasner 1984). Continuity, a chief concern of historical institutionalism, is maintained through feedback mechanisms (Pierson 1993, 2004). Feedback mechanisms work through the distributional effects of institutions, making institutions central in 'reflecting, reproducing and magnifying existing patterns of power distribution' (Thelen 1999: 394). As Thelen notes, this approach includes two related but analytically distinct claims. The first involves arguments about crucial founding moments of institutional formation that send states along different developmental paths; the second, that institutions continue to evolve in response to changing environmental conditions and ongoing political manoeuvring, but in ways that are constrained by past trajectories (Thelen 1999: 387).

The relevance of these insights for party development is obvious. Party politics literature has long stressed the lasting effect of origins for subsequent party development, identifying a similar kind of path-dependent dynamic (Duverger 1964; Panebianco 1988). Panebianco comments directly on this process: 'The way in which the cards are dealt out and the outcomes of the different rounds played out in the formative phases of an organisation, continue in many ways to condition the life of the organisation, even decades afterwards' (Panebianco 1988: xiii).

United Russia, as detailed in Chapter 2, is part of the same party strategy that has persisted throughout the post-Soviet and even late Soviet period. As a result, United Russia is subject to a range of path-dependent effects, while its persistence and evolution have been shaped by feedback mechanisms that incorporate the idea of elite learning, of actors attempting to reach equilibrium with their political environment. This research begins by placing United Russia in context, tracing its origins to the Yeltsin period, first with the pro-presidential Unity movement in 1999, but importantly to the Our Home is Russia movement in 1995, which in many ways was the prototype for United Russia in the post-Yeltsin period. This historical perspective is essential for understanding United Russia's development, role and relationship with dominant-power politics.

Theoretical framework

In theoretical terms, the aim of this study is to identify the mechanisms that link United Russia with dominant-power politics in the post-Yeltsin period, 2000–10, mechanisms that also enable an understanding of its role. This framework is not intended to predict or model a regular succession of events, but to guide the empirical investigation, to set up a standard ruling party–regime outcome model derived from existing literature and then to look at power perpetuation in more detail. The underlying comparative or 'portable' mechanism (Falleti and Lynch 2009) is that organisations use power to consolidate and gain more power. This is also expressed with the idea of a snowball or bandwagon effect, in which one party, gaining a clear advantage over other parties, is able to 'use that advantage to reinforce its dominance' (Rasmussen 1969: 407), in this case, that ruling parties use their organisation and power to generate dominant-power politics. However, this study is not concerned with validating or disproving this point. Instead, the emphasis is on identifying when or under what conditions this happens and identifying the effects of context on this process.

Elster (1989: 4–8) defines a (causal) mechanism as part of a detailed explanation of causality that is neither a simple citation of cause, nor an identification of correlation, necessity, storytelling or prediction. This definition differs from those that identify mechanisms with elements of law-like regularity (Little 1991: 15). Law-like regularity is a feature of the natural sciences, but in the open systems of the social world, several mechanisms may work together, serving to mitigate each other's effects. The stratified ontology of social reality means that the levels of the actual, empirical and real combine to limit the possibility of law-like regularity (Bhaskar 1978). Indeed, it is the importance of context, highlighted by Falleti and Lynch (2009) and others, that this study is keen to explore. As noted, history matters, so the ability of United Russia to contribute to dominant-power politics in the same way as other ruling parties elsewhere depends to a great extent on the circumstances of Russia's overall emergence from the Soviet Union. The mechanisms identified in the comparative literature are mitigated by context, in this case the conditions that (still) serve to limit the overall power of parties, including the ruling United Russia.

As this research deals with party evolution over time, of its development into a more complex form, the subject of social change deserves mention. The approach that this study takes to structure and agency, in line with Hay (1996, 2002) and Jessop (1990), is to treat both as relational and dialectical, neither separate nor the same, but interwoven (Hay 2002: 127). As already mentioned, United Russia is a strategy conditioned by its political environment, but an environment that is changing, in turn effecting the calculations of actors. This environment is also altered by active strategies of learning; what may be called the strategic calculation of actors regarding their situation (Hay 1996: 124). Overall, whether it is the strategic calculation of actors, elite learning or the effects of agency, the 'elite variable', as it is termed in Chapter 2, is crucial for understanding United Russia, its relationship with dominant-power politics and its role in the post-Yeltsin regime.

Empirical perspective

Over 80 face-to-face, in-depth elite interviews were employed as the primary data-gathering method for this study, although a range of primary and secondary materials were also utilised. The term 'elite' deserves mention here as it is used frequently in this study, but remains open to varying definitions in the literature. The treatment of elite typically refers to those individuals who hold a privileged position in society (Richards 1996: 199) or, more specifically, who control a disproportionately high level of political, social and economic resources, 'giving them the capacity to make decisions binding in the larger community' (Waldner 1999: 22 fn10). However, this research uses two definitions of elite interchangeably. In Chapter 2, the term elite variable is introduced, referring to those persons able, because of their strategic positions in powerful organisations, to affect national political outcomes regularly and substantially (Burton *et al.* 1992: 8). In addition, and for the purposes of fieldwork, the definition of elite was much broader, understood to mean a crucial group for understanding a particular sector of society as a whole. In this sense, all elites interviewed for this research can be considered experts and all the experts considered elites; the important point is that they had information on United Russia and dominant-power politics. This elaboration of elite supports the idea of elites as informants/experts about a specific field of investigation (Hoffmann-Lange 1987: 28).

The site chosen for the interviews was Moscow; a practical choice considering the size of the Russian Federation and the time and financial limitations that every research project encounters. Moscow, as the political and economic centre of the country, is a good site to access elites and has the added advantage of a large and varied expert community. The State Duma and Federation Council also offered the chance to interview regional elites. Aside from allowing access to the party-based elites that formed the crucial group of this research, Moscow City, as a subject of the Russian Federation, also has its own regional legislature, meaning that interviews were conducted at every level of governance, including municipal, regional and federal levels.

The significant difficulties encountered while conducting fieldwork in Russia also justified a prolonged stay in Moscow. Research spanned a period of eight months, divided into two periods. The first period ran from February to August 2007, with a second period of follow-up research conducted in October and November 2007. This follow-up research also coincided with the campaign period for the December 2007 State Duma election, which enabled the first-hand observation of an election campaign in process. In total, 85 interviews were conducted during the fieldwork period, including 33 interviews with party elites and 19 interviews with elites from other organisations. The remaining 33 interviews were conducted with experts, including several journalists. Overall, this research benefited from the input of many eminent political figures, including, and in no particular order, former Russia's Choice leader, Egor Gaidar; *Izvestiya* editor, Vladimir Mamontov; former Our Home is Russia leader, Vladimir Ryzhkov; former Unity leader, Alexsandr Gurov; former URF leaders, Nikita Belykh and Boris Nemtsov; Yabloko leader, Sergei Mitrokhin; United Russia chair of Central Executive Committee, Andrei Vorob'ev; government representative in the State Duma, Andrei Loginov; president of the Polity Foundation, Vyacheslav Nikonov; head of the INDEM Foundation, Georgii Satarov; director of the Centre of Political Technologies, Igor Bunin; head of the Globalisation Institute, Boris Kagarlitskii; Moscow Carnegie Centre scholars Andrei Ryabov, Nikolai Petrov and Aleksandr Kynev; director of the Panorama think tank, Vladimir Prybilovskii; and many others. The interviewees cited in the chapters that follow state their occupation or official capacity at the time of the interview, although in an effort to maintain ethical standards some interviewee details are omitted altogether.

In addition to these interviews, data gathering also included numerous unrecorded observations from round tables and informal discussions, as well as secondary sources such as official statistics, archival material, biographical data, data from party publications, data from organisations and associations linked to parties, opinion poll data, newspaper articles and academic publications. Despite the evident strengths of elite interviewing, the drawbacks of this method can be reduced, essentially, to problems with validity and reliability (Davies 2001; May 2001: 141–4). In the case of elite interviews, the problem of reliability is clear enough to warrant triangulation whenever possible (Lilleker 2003) and this also justifies the relatively large number of expert interviews included in this research. In addition, efforts were made to tally information given in interviews with secondary sources, such as material in the Russian media; a task greatly aided by the accessibility of archived newspaper reports on the internet. The internet also proved a valuable source of data, especially official government and party websites.

Structure of the book

Although United Russia's relationship with dominant-power politics and the party's overall role in the politics of the post-Yeltsin period form the primary foci of this study, the following six chapters also explore a number of secondary issues, including how to understand the party's rapid rise and the problematic party of power (*partiya vlasti*) term. Chapter 1 presents the theoretical framework that provides a heuristic ruling party–dominant-power politics model or ideal-type relationship, highlighting three major areas or party roles, including managing elections; governing and enforcing incumbent authority; and integrating the elite and society. Chapter 1 also discusses the nature of dominant-power politics in Russia, before introducing the principal–agent distinction as the basis for the substantive argument of this study.

Chapter 2 develops this argument further by identifying the antecedents of United Russia and the party of power in general, showing how context, seen through the specific detail of party creation, matters for understanding party roles and regime outcomes. Aside from identifying parties of power more clearly as party agents, this chapter shows how party development in the early 1990s continues to affect United Russia at the end of 2010. This chapter also provides a preliminary answer to the question of how United Russia rose to prominence in the first place by showing how ruling parties also reflect as well as affect larger, macro-political outcomes.

The remaining chapters examine United Russia in line with the theoretical framework outlined in Chapter 2. Chapter 3 deals with elections in the post-Yeltsin period, Chapter 4 with governance and Chapter 5 with integration. Each chapter provides a focused exploration of the contribution of the party to dominant-power politics within these areas. Chapter 6 summarises the material presented and returns to the main questions of this study.

1 Ruling parties and dominant-power politics

The theoretical framework

We have a one-party dominant system, which I think is very healthy in Russia. In a situation where Putin is definitely more liberal than 90 plus per cent of Russians, that is the only way to pursue any reforms in this country, which is completely resistant and still communist and nationalistic in nature.

(Vyacheslav Nikonov 2007)

I can see two roles for them [parties of power]. The first one is the role of clientele . . . the second role is to imitate a real and active party-system.

(Georgii Satarov 2007)

This chapter distils existing literature to provide what is, in many ways, a standard framework to study United Russia and other dominant ruling parties. As such, the major contribution that this chapter makes is found not in answer to the question of *how* parties generate dominant-power politics (although this is reconsidered in the conclusion chapter), but of *when* and under what conditions parties appear as major explanatory variables in a range of macro-political outcomes, including dominant-power politics. From a comparative perspective, ruling parties operating in very different political systems share many similarities, but they also have unique characteristics relating to the general and specific context in which they are found. For this reason it is important to consider the mechanisms of dominance, as only then is it possible to fully appreciate issues of power and stability, as well as the role of parties.

As noted in the introduction chapter, United Russia's electoral, parliamentary and organisational dominance, as well as the weight of evidence from comparative literature, strongly suggest that the party is indeed an explanatory variable in the larger outcome of dominant-power politics in the post-Yeltsin period. However, United Russia represents a puzzle in the sense that Russian parties in the 1990s, with the exception of the CPRF, were weakly articulated and largely peripheral to political processes occurring in the country. In order to resolve this puzzle, it is important to move beyond purely quantitative indicators of party strength, such as organisational extensiveness and electoral success, to consider ruling parties and their role in more detail.

The theoretical framework outlined in this chapter draws on a number of studies dealing with the role of parties in authoritarian regimes, in particular the work of Brooker (1995), Brownlee (2007), Way (2006) and Levitsky and Way (2010). All four engage with parties in non-democracies and each support the centrality of ruling parties as explanatory variables in regime outcomes. Way (2006) and Levitsky and Way (2010) and their work on competitive authoritarianism identify ruling parties as important components of incumbent capacity, essential for understanding authoritarian persistence and breakdown in the post-Cold War period. According to Way (2006: 15–17), ruling parties play an important role in uniting elites, managing the electoral process, controlling the legislature and facilitating executive succession – all to the benefit of authoritarian regime persistence. Brooker (1995: 16–20), in his earlier study of twentieth-century dictatorships, also identifies political parties as key components of authoritarian power, ascribing them political, governing and social roles important for maintaining the regime. This study employs a similar framework for understanding dominant-power politics in Russia. United Russia is explored in terms of its electioneering, governing and integrating roles, addressed in Chapters 3–5.

Russia as an exceptional democracy?

The discussion that follows is wide-ranging, detailing both the theoretical framework and its practical application. The second part of this chapter considers the mechanisms through which ruling parties generate dominant-power politics. The final part of this chapter considers when these mechanisms are activated. However, the starting point is the dependent variable itself, in this case dominant-power politics, its nature and the post-Yeltsin regime in general.

Dominant-power politics and Russia's democratic credentials

In view of the large and established literature dealing with the role of parties in democracies, the first task that must be resolved is whether this literature and these insights are suitable for analysing the Russian case. Are there grounds for viewing dominant-power politics in the post-Yeltsin period (2000–10) as another example of what Pempel (1990a) terms 'exceptional democracy', in which a ruling party, in this case United Russia, defies the odds of competitive, multi-party elections to retain power over the course of several election cycles? What is the justification for not employing a normative, democratic analysis of United Russia and its role in securing longer-term democratic consolidation?

The first point to make is that, although an analysis of United Russia's potential to oversee democratic consolidation is possible, from a theoretical and empirical standpoint it would be both far-sighted and ambitious. From a purely theoretical perspective, the application of democratic theory to the Russian case is by no means straightforward, especially when so much within this broad body of literature is based on the experiences of the first wave (1828–1926) and second wave (1943–62) of democratic transitions and when the results of the most recent third wave (1974–97) suggest a different party-democracy dynamic (see Lewis 2001).[1] The fact that democratic theory emerged after the appearance of parties, not before (Crotty 2006: 25), may actually serve to ground it in the experience of Western European/North American political development, where parties did play an important role in creating and supporting democracy. We should not expect the same linear relationship between parties and democracy in different temporal and spatial contexts, based on this fairly narrow historical experience.

In the first wave of democratic transition, the early American political parties played a significant role in overcoming the difficulties surrounding nation-building and in helping to institutionalise the nascent democratic system (Chambers 1966: 90).[2] Parties played a no less important role in the development of democracy in Western Europe (Epstein 1980; Rokkan 1975), appearing at the crucial stage of state-building when the masses began to actively participate in the political system. In the European context, parties were instrumental in mobilising support, and articulating and aggregating demands, to the benefit of democracy as a whole (Rokkan 1975: 570). Like the first wave of democratic transition, the second wave also saw parties figure prominently, especially in post-war Japan, West Germany and Italy, although by the end of the second wave the relationship between parties and democratic outcomes was not so consistent. Parties were central to political development in much of post-colonial Africa and Asia, but not always to democratic ends.[3]

Overall, the relevance of existing democracy-centred theory depends largely on the degree to which these theories are generalisable from earlier waves to later waves. The three waves, at face value, are very different (Huntington 1991–2: 580) and there are grounds to question the relevance of North American and European historical experience for understanding later democratic waves (Bunce 1995; Mainwaring 1999: 21; Mainwaring and Torcal 2005: 1; Munck 2001: 121; Rose and Shin 2001; Tilly 1975: 601).

In summary, there is no question that United Russia, with its large membership, all-national organisation and dominant position in the party-system, appears well placed to play a central role in overcoming the problems that undoubtedly exist in the Russian political system, and to institutionalise the party-system and consolidate democracy. But there is no compelling evidence that Russian democratic development will repeat the first or second wave experiences. At the same time, by labelling post-Soviet democratisation a separate 'fourth wave' (McFaul 2002) some scholars already acknowledge that Russia's multiple transition following the collapse of the Soviet Union shares little in common with even the most recent third wave. This has obvious implications, as even with the substantial conceptual stretching required to formulate dominant-power politics in Russia into purely democratic terms there is no guarantee that existing democracy-centred theory will be of any real use for answering the questions identified at the beginning of this study.

The Russian political system for much of the post-Soviet period falls into what has been termed a 'grey zone'; a mixture of democratic and authoritarian features that make regime classification difficult. Grey zone is attributed to Carothers

(2002) and his critique of the transition paradigm, although many authors have applied this term to post-Soviet Russian politics (Colton and McFaul 2003: 13; Gel'man 2005; Ledeneva 2006: 29). Grey zones arise for a number of reasons, undoubtedly reflecting an analytical deficiency in classifying regimes that occupy the space between two ideal types: those regimes that are completely open and those that are completely closed.

When it comes to applying theoretical insights, including those from democratic and transition theory, the nature of Russia's grey zone as either more or less democratic is of great importance. As Colton and McFaul note, 'if Russia is a democracy, then theories that explain transitions to democracy may provide a meaningful framework for understanding regime change in Russia' (Colton and McFaul 2003: 13). It goes without saying that if Russia is neither in a transition to nor consolidating democracy, then democracy as a whole is not an appropriate system-level dependent variable for this study.

One of the problems with accurately gauging the democratic credentials of Russia or any other state is that, in the absence of any legitimate alternative, democracy remains the master text to which most states (at least in an official discursive capacity) subscribe. As such, it is no real surprise that every post-Soviet state, from Belarus to Turkmenistan, is defined by its own constitution as democratic. This means that penetrating the grey zone is made more difficult by the lengths that many states are prepared to travel to create a democratic façade with little content or substance (see Wilson 2005). This problem is exacerbated by either the perception or the fact that democratic standards are sometimes applied unevenly and used as political leverage by the West against other states. The result is that democracy remains a highly contested concept, and so the belief that empirical investigation is all that is required to penetrate the grey zone and provide clear and consensual lines between regime types (Schedler 2002: 38) underestimates the inherent problems of any classification scheme.

In the case of post-Soviet Russia, political leaders from Yeltsin to Medvedev have consistently supplied a democratic discourse for domestic and foreign audiences. Vladimir Putin's keynote presidential addresses, 2000–7, are a case in point, containing consistent reference to the importance of democratic development in Russia, indicating that Russia's rulers broadly accept the principle if not the degree of pluralism inherited from the Yeltsin period. During Putin's second term of office, 2004–8, the Russian leadership reconfigured this discourse allowing the emergence of so-called 'sovereign democracy', a theme examined in more detail in Chapter 5, but adding to the confusion surrounding the nature of dominant-power politics in Russia.

In general, Russia's status as a democracy in the post-Yeltsin period is not without any support from Western observers, although supporters are few and have become fewer. If during Vladimir Putin's first term of presidential office (2000–4) it was possible to find some grounds for optimism in assessments of Russian democracy (Bunce 2003; Shleifer and Treisman 2004), by the beginning of Putin's second term of office (2004–8) more pessimistic appraisals predominated (Fish 2005; Ledeneva 2006; Lynch 2005; Wilson 2005).

This dichotomy of optimists and pessimists was certainly clearer in the early years of Putin's first term of office. Smyth, for example, commenting on the overall approach to Russian politics, refers to pessimistic scholars who 'leave normal politics out of their assessments . . . focusing only on the extra constitutional or coercive aspects of the Putin regime' (Smyth 2002: 573). Hanson also identifies 'pessimistic scholars' who conclude that the Russian Federation is no more than a democratic façade, based on institutional mimicry of the West (Hanson 2001: 127). However, the pessimists were certainly in the ascendancy by the end of Putin's second term of office and little has changed in this regard by the end of 2010.

In reality, support for Russia as a democracy has always been qualified, even in the Yeltsin period. After the initial democratisation following the collapse of the Soviet Union, 'delegative' democracy became the first in a series of adjectives used to place a limit on democracy in Russia. Delegative democracies are characterised by the premise that 'whoever wins election to the presidency is thereby entitled to govern as he or she sees fit, constrained only by the hard facts of existing power relations and by a constitutionally limited term of office' (O'Donnell 1994: 59). The establishment of the 1993 Constitution of the Russian Federation and the creation of super-presidentialism (Colton 1995) endowed the institution of the presidency with wide-ranging powers that, although justified on the basis of the political and economic uncertainty of the time, also created substantial risks for democratic development (Kubicek 1994: 435). It is interesting to note that even the democratic constitution adopted in December 1993 (a replacement for the 1977 Soviet Constitution) was considered by some Russian analysts to be rather undemocratic (Nikonov 1993).

The same kind of qualified democracy can be seen in the literature at the end of the 1990s and into the early post-Yeltsin period, framed in terms of 'protracted transition' to democracy (McFaul 1999) or 'managed democracy' (Colton and McFaul 2003). Colton and McFaul used managed democracy to acknowledge the democratic gains made in the Yeltsin period, but also the erosion of these gains in the post-Yeltsin period, avoiding placing Russia into a non-democratic category in the process (Colton and McFaul 2003: 14). Hanson, another to frame Russia's political system in terms of democracy, justifies this approach by highlighting the comparative advantage of using existing democratic-transition stages as a framework for analysis, identifying Russia as an example of 'unconsolidated democracy' (Hanson 2001: 128). Aside from delegative, managed, unconsolidated democracy and protracted transition, some of the many other Yeltsin and post-Yeltsin labels include 'regime politics' (Sakwa 1997), 'managed pluralism' (Balzer 2003) and 'poor democracy' (Aron 2007: 203–15).

The normative implications of democracy and authoritarianism in the post-Cold War period and the short time frames available to gauge post-Soviet political development make the question of Russian democracy impossible to answer in any satisfactory way. The independent watchdog organisation in support of freedom, Freedom House, which measures freedom on a scale of one to seven, with one representing the highest degree of freedom, shows persistent backsliding in Russia throughout the post-Soviet period, suggesting, if anything, a transition away from democracy (see Figure 1.1).[4]

Although the freedoms measured by the Freedom House scores are by no means universally accepted (nor the independence of this organisation), they do feed into the practical functioning of free and fair elections, which in turn offers a clearer procedural understanding of dominant-power politics in the Yeltsin period. The Russian political system, like any democratic system, relies on popular elections and voter choice as the primary means for resolving power issues and for providing rulers with legitimacy. Article 3 of the Constitution of the Russian Federation (CEC 1993a) confirms the place of free elections as the 'supreme direct expression of the power of the people'.

However, in the post-Yeltsin period, there is no shortage of evidence that shows Russia to be far from its own constitutional ideal regarding the electoral process. Although the constitution contains no reference to fair elections, it is difficult to ascribe free elections with any meaning unless a degree of fairness is introduced into the equation. Ledeneva (2006) and Wilson (2005), for example, cite several persistently unfair features of Russian elections, including manipulative campaigns, the prevalence of black PR and the use of administrative resources to tilt the electoral playing field in favour of incumbents, what Wilson terms the 'abuse of the state's administrative resources to defraud the electoral process' (Wilson 2005: 73). Although the use of manipulative election campaigns and black PR is not confined to Russia alone, Ledeneva identifies several characteristics of the

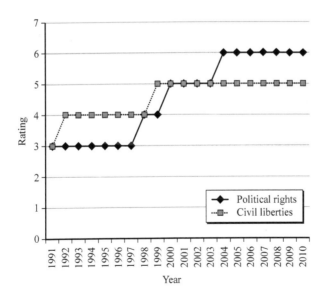

Figure 1.1 Post-Soviet Russia: Freedom House scores. Source: Freedom House (2011).

Russian electoral process that, taken together, makes it slightly more exceptional. These include widespread manipulation of the law, less constraint on election financing, the use of journalism on demand to facilitate black PR and a backdrop of political volatility that provides perfect conditions to make and break political reputations and to influence swing voters (Ledeneva 2006: 45). Fish (2005: 30–81), among others, notes several major shortcomings of the Russian electoral process, including:

- ballot stuffing;
- buying votes from voters;
- pressure on state workers, including the military, to vote for incumbents;
- soft coercion or intimidation of voters;
- hard coercion or political violence against opposition;
- the arbitrary exclusion of candidates from elections;
- unequal media access and media coverage of elections.

These examples are by no means exhaustive, but are corroborated by the Office for Democratic Institutions and Human Rights (ODIHR) and by a raft of media accounts. Media reports on election infringements in Russia are so numerous that it is difficult to know where to begin when documenting them, although it is worth noting that the useful overview of Russian elections given by Fish (2005) is based primarily on reports from *The Moscow Times*. The ODIHR monitored several Russian elections, 2000–4, but not the 2007/8 election cycle, because of disagreements with the Russian authorities. Nonetheless, the ODIHR did have serious concerns regarding the elections that were monitored, including their overall lack of fairness (ODIHR 2004).

Moreover, there is little indication that these violations have abated with the arrival of Dmitri Medvedev as the new president of the Russian Federation. Among the many reports of inconsistencies in the period 2008–10, the October 2009 regional parliamentary elections are particularly indicative of the continued weakness of Russian electoral democracy. On 14 October 2009, State Duma deputies from the opposition Liberal Democratic Party of Russia (LDPR), CPRF and A Just Russia staged an unprecedented walkout in protest of what they considered widespread falsification in favour of United Russia and the use of administrative pressure on candidates and voters in a number of regions (Rodin and Samarina 2009a). The Yabloko party's website documents a range of allegations of electoral fraud in Russia, which, from the perspective of this opposition party at least, shows that elections by the end of 2010 are becoming less fair as the regime struggles with the effect of the 2008 financial crisis while attempting to further consolidate its power.

Elections in many countries are subject to sub-democratic practices, but no electoral democracy can support such systematic violations of the electoral process, even if completely free and fair elections are an unattainable ideal. Elections, as a crucial minimum criterion, suggest that the Russian democratic glass, 2000–10, is half empty rather than half full.

Defining dominant-power politics

The course of post-Soviet Russian politics, from the Yeltsin to post-Yeltsin periods, appears to be a transition from instability to stability. As mentioned in the introduction chapter, feckless pluralism and dominant-power politics were originally employed by Carothers (2002) to describe the two major tendencies that he perceived within the third wave of democratic transition. Feckless pluralism describes those states where democracy remains shallow and troubled, while dominant-power politics identifies states where, despite democratic institutions, one political grouping dominates, undermining any sense of power contestation and power alternation (Carothers 2002: 9–14).

Dominant-power politics in Russia, 2000–10, in more concrete terms, resembles what has elsewhere been termed electoral authoritarianism (Golosov 2008; Schedler 2002, 2006a). An electoral authoritarian system is one in which elections hold a great deal of significance even when subject to widespread manipulation in favour of power-holders. Their importance involves the claim that 'the intrinsic power of elections, more than anything else, drives the dynamic of stability and change' (Schedler 2006b: 12). Schedler considers any regime with a combined average Freedom House score of four or above (political rights and civil liberties) to be electoral authoritarian and any regime with a score of below four to be electoral democratic (Schedler 2002: 47). Lindberg makes a three-way distinction between liberal democracy, scoring a maximum average of 2 on the Freedom House score; electoral democracy, scoring higher than 2 but less than 3.5; and electoral authoritarianism, scoring 3.5 or more (Lindberg 2006: 151–2). As of 2010, Russia has a combined average rating of 5.5, placing its democracy on a level footing with Algeria, the Republic of Congo and Iraq (Freedom House 2010), while *The Economist* Intelligent Unit (*The Economist* 2010) assigns Russia an overall rank of 107 out of 167 countries on its Democracy Index.

Importantly, and in addition to unfair elections, other aspects of post-Yeltsin politics reinforce the characterisation of dominant-power politics as a variety of electoral authoritarianism rather than electoral democracy. D'Anieri (2007: 216), for example, associates electoral authoritarianism with the following tactics designed to allow incumbents to avoid the major uncertainties of democratic governance and political completion:

- selective law-enforcement;
- selective administration of regulations;
- control over the media;
- control over patronage;
- control over the economy.

D'Anieri also ascribes a broader, post-Soviet institutional context to electoral authoritarianism that includes extensive executive power, a weak parliament and a weak judicial branch (D'Anieri 2007: 216). Russia in 2010, as the following chapters testify, qualifies on all counts.

Electoral authoritarianism is part of the recent attempt to define the grey zone in the belief that both the transition paradigm and existing authoritarian typologies are no longer useful. This has also been accompanied by a move away from qualitative regime typologies towards quantitative measures or scales that grade regimes according to the competitiveness of their elections and openness of their societies (Collier and Adcock 1999; Fish 2005; Munck 2006).

There is some basis in the notion that some qualitative typologies are inadequate in the face of this present democratisation dynamic, in which a growing number of regimes attempt to gain democratic legitimacy without tolerating the uncertainty that democratic elections entail. Classic typologies of non-democratic regimes were primarily based on a distinction between totalitarianism and authoritarianism, but this typology soon grew obsolete with the realisation that few regimes approached the totalitarian ideal type, making the authoritarian type too inclusive (Hadenius and Teorell 2007: 144).

In this respect, electoral authoritarianism is not a new occurrence, as the use of adjective authoritarianism is clearly evident in classic studies such as O'Donnell's account of 'bureaucratic authoritarianism' (O'Donnell 1973). It should also be noted that authoritarian regimes that gain their legitimacy through multi-party elections contested by opposition parties, as shown in the next section, predate the present-day Russian case by some time. As mentioned in another classic work, authoritarian regimes are characterised by a pluralistic element, but a limited pluralism rather than the unlimited pluralism of democracies (Linz 1970: 255). This fits well with the recent work on so-called competitive authoritarianism (Levitsky and Way 2010) as well as the aforementioned electoral authoritarianism.

Understanding the post-Yeltsin period in terms of a transition from feckless pluralism and weak electoral democracy to dominant-power politics and electoral authoritarianism does not make any definitive statement on Russia's potential for future democratic development. Nor does it rule out the longer-term benefits of this transition for the Russian state, as the quote from Vyacheslav Nikonov at the beginning of this chapter indicates. At present, it is not suitable to approach dominant-power politics from the perspective of a normative or democracy-centred analysis. Looking at the role of United Russia in terms of its actual or potential effect on democratic development simply raises the prospects of concluding that the party plays no role at all, and falls into the unfortunate normative trap of applying a transition approach to Russian politics, despite contrary evidence.

The mechanisms of party dominance

The framework used in this study highlights three broad party roles – managing elections, governing and integrating – which taken together encompass the mechanisms through which ruling parties generate dominant-power politics. This framework is supported by existing literature on the role of parties in non-democracies, in particular the work of Geddes (2006) and Way (2005), together with a range of single studies of ruling parties and dominant-power politics from a number of countries, mostly from the twentieth century. Essentially, what proceeds is a characterisation of an ideal-type ruling party drawing on a range of comparative examples, including the Mexican PRI, the United Malays National Organisation (UMNO) and the Taiwanese *Kuomintang* (KMT), but also single-party regimes (CPSU) and democratic regimes (Japanese Liberal Democratic Party, LDP) to illustrate these roles and the mechanisms of dominance effectively. In sum, ruling parties generate dominant-power politics by controlling the electoral process, by governing the polity in the interests of the ruling party and by effectively integrating elites and society to stabilise the regime.

The electoral process and ruling parties – managing elections to perpetuate power

Although there is a great deal of variation in electoral systems in operation around the world, their degree of transparency and the extent to which elections are symbolic as opposed to real arenas for power contestation, the number of closed regimes, where elections are non-existent, are few. By 2010, and according to data from the American Central Intelligence Agency (CIA 2010), only Brunei Darussalam, Eritrea, Saudi Arabia, Somalia and Qatar exist as major, independent states that have not held any form of national legislative election within the past 20 years, although a number of states do elect legislatures indirectly or without universal suffrage or minus the participation of opposition parties/candidates (at least, not independent of the ruling party). They include China, Cuba, Laos, Libya, North Korea, Oman, Turkmenistan, United Arab Emirates and Vietnam.

This means that, in the overwhelming majority of states, ruling parties must mobilise voters as the essential requisite for retaining power, but the way that this is done will obviously depend on the effectiveness of the rule of law and the competitiveness of the electoral and political system in question. As Schedler comments, the 'dream of autocrats is to reap the fruits of electoral legitimacy without running the risks of democratic uncertainty' (Schedler 2002: 37). As such, the task of the ruling party in semi-competitive systems is often to manage elections in order to facilitate favourable results, rather than facilitate any genuine participation and competition to anything approaching a democratic norm.

In the case of Russia, 2010, control over the electoral process is strong, but by no means total. Although the transition from feckless pluralism to dominant-power politics suggests increasing levels of incumbent control – a suggestion that is not disputed in this study – the logistics of electioneering or even coordinating electoral fraud in a country the size of the Russian Federation always leaves room for error; a subject given more detail in Chapters 3 and 5. In any semi-competitive political system, the task of the ruling party is by no means easy. The party must secure election victories, but in a way that preserves the integrity of the electoral process as a whole. Elections provide a great deal of legitimacy for power-holders, but only when blatant fraud does not undermine the result.

At the same time, for elections to be meaningful they require a degree of opposition and competition to satisfy voters. Although complicating the task of the ruling party, the presence of opposition has become increasingly essential to meet nominal international standards too. Overall, elections won against opposition parties afford a great deal more legitimacy to power-holders than plebiscitary elections and have the advantage of performing two dominant power-sustaining functions. Allowing opposition to compete in elections and even gain some limited success can provide a safety valve to release political pressure by allowing opponents and dissatisfied voters the opportunity to vent their frustrations with the regime. In addition, inflated winning margins contribute to an 'image of invincibility' for the ruling party, which, when combined with some small success for opposition parties, may be enough to deter future regime challenges (Geddes 2006; Magaloni 2006: 234). It is no surprise then that semi-competitive elections, biased in favour of the ruling party, are what Brooker terms 'the most sophisticated type of undemocratic election' (Brooker 2000: 107). The bottom line is that elections, regardless of the degree of incumbent control over political processes, always offer some opportunities for opposition forces, reinforcing the role and importance of the ruling party to skilfully manage them.

Overall, when exploring this electoral aspect of dominant-power politics in more detail, there are a number of common denominators found in comparative literature. The first point to note is that many ruling parties are well funded and well organised, employing talented staff, enjoying access to significant state resources, notably state-controlled media. This forms part of what Greene (2007) terms a 'resource theory' of single-party dominance. In the case of Mexico under the PRI, for example, dominant-power politics owed much to the party's financial and media dominance and so it is of no real surprise that the party's eventual loss of electoral dominance from the end of the 1980s owed much to the increased media and financial access afforded to opposition parties (Brinegar *et al.* 2006: 82).

This, arguably, forms part of the more traditional party electioneering role, of deploying resources and communicating with and mobilising the electorate – an aspect that may not differ greatly from the role of ruling parties in maintaining dominance in democratic systems. The Taiwanese KMT, for example, literally 'learned how to win elections' after gradually allowing opposition parties to form and compete in elections from the 1970s onwards (Cooper 2003: 147–8). Professional staff and effective media usage, in particular the KMT's success in characterising the opposition Democratic Progressive Party as likely to bring war with China and ruin to Taiwan, helped the party maintain its dominance, even with increasing levels of electoral competition.

In a similar way Rasmussen (1969: 409) notes that the United National Independence Party (UNIP) quickly became the dominant party in Zambia in the 1960s because of 'the attractiveness of its programme, its organisational effectiveness and the quality of its leadership'. This stresses the importance of understanding party dominance in any political system as a function of the ruling party's interaction with the electorate, rather than of rigged elections. Mair (1979: 458) summarises this dynamic well with the observation that the virtuous cycle that saw Fianna Fáil dominate government in Ireland from the early 1930s owed as much to the party's ability to appeal to a political culture with a predilection for strong leadership and to benefit from the tendency for Irish voters to support the party most likely to win.

Ruling parties enable an organised and coordinated election effort, mobilising voters through a range of means, from party patron–client networks to door-to-door campaigning, in order to secure the right results for the regime (Way 2006: 16).[5] Strong ruling parties with extensive party networks are also effective in distributing state resources to supporters between election campaigns. This helps to build a durable base of voter support and reduces the need for blatant fraud and high levels of coercion during election campaigns, simply because party machines, compared with their competitors, are better placed to succeed. In Egypt, for example, the fact that state resources are almost entirely restricted to candidates of the ruling National Democratic Party (NDP) in no small way explains the party's successive parliamentary majorities and decades of dominance (Albrecht 2007: 64 fn8) – at least before the mass actions and regime upheaval of early 2011. This favourable resource environment provided NDP candidates with a considerable advantage over competitors and removed the need for excessive fraud or coercion simply because the ruling party machine was better placed to succeed.

In practical terms, ruling parties contribute to dominant power at election time not only by delivering victories, but also by delivering the required levels of ruling party representation in the institutions of power, which then allows incumbents to further consolidate their position. The task of the ruling party is to win elections, but also gain enough votes to ensure control over the executive and legislative branches. Control over the legislature is important for a number of reasons. Legislative majorities enable incumbents to make far-reaching reforms that are legitimated by the electoral process. The larger the majority, the less opposition forces are able to scrutinise these reforms. It has been noted, for example, that the PRI in Mexico invested heavily in 'super-majorities' during parliamentary elections in order to make changes to the constitution. Constitutional changes inevitably contributed to the perpetuation of the PRI hegemony; the constitution changed almost 400 times to the advantage of power-holders, including changes to electoral institutions and the weakening of judicial power (Magaloni 2006: 259).

The governing role – ruling in the party's interests

Parties in a democracy govern according to interests, often broad societal interests, and this forms the theoretical mechanism that makes representative democracy possible (Downs 1957; Katz 1997; Riker 1988). Ruling parties also contribute to dominant-power politics by governing, but here the emphasis is on representing the regime's interests in policy-making, to establish and consolidate control. As noted previously, in many states with directly elected parliaments, elections usually determine the extent of the ruling party presence in, and control over, executive and legislative branches. In turn, this control increases the possibilities for the ruling party to streamline legislative–executive coordination with a limited presence of opposition veto players. The ruling party may use its dominance over the formal institutions of governance to pass law to further centralise and concentrate power, provide pay-offs to allies or simply limit the threat of opposition through targeted legislation. Ruling parties in control of both executive and legislative branches create the perfect conditions for the virtuous circle of dominance, where control begets greater control.

In Malaysia, for example, control over executive and legislative branches enabled gerrymandering of electoral constituencies in favour of the ruling UMNO, turning the marginal ethnic Malay majority into a potential two-third or greater parliamentary majority (Brooker 2000: 246). In a similar way, UMNO was able to pass wide-ranging security legislation that served to limit the effectiveness of opposition (Case 1994: 918). In some cases, ruling party control over the institutions of governance allows the passage of laws directed at controlling potential opposition, such as independent media. In Yemen, for example, the ruling General People's Congress was able to pass a comprehensive press law that included requirements for journalist qualifications, newspaper registration, publishing permits and conditions for press closures, as well as the admittance of foreign journalists and publications (Lawson 2007: 123). These laws strengthen the ruling party by limiting potential outlets for opposition voice.

Control over the legislative process also creates the opportunity to use economic levers to channel resources and buy support at election time. This has been identified as an important aspect of dominant-power politics in a number of African states, where public expenditures targeting specific constituencies serve to maintain cohesion among allies as well as boost incumbent popularity among the electorate as a whole (Bratton and van de Walle 1997: 75). Often, the 'self-enforcing hegemony' of the ruling party depends on the support of the larger public and this is secured largely by controlling the central government and fiscal system (Diaz-Cayeros *et al.* 2003: 36). Control over the fiscal system may simply translate into the ability of the ruling party to increase pensions or provide other monetary 'sweeteners' at crucial times, such as before elections or when popularity is flagging.

In this respect, there is little difference between dominant-power politics in electoral authoritarian regimes and parties in democracies. Office-holding parties in democracies often use well-timed budgets and a host of institutional advantages to increase their prospects of re-election – and even to deter competition (Katz and Mair 1995). Examples of patronage politics to secure support can be found in many electoral democracies. The ruling Japanese LDP, whose long-term dominance is often viewed as one of the few cases of perpetual, single-party dominance among democracies (Maramatsu and Krauss 1990: 282), traditionally enjoyed close connections with bureaucracy and big business forming an 'iron triangle' or tripartite alliance of interests (Bowen 2003: 86). Between August 1990 and December 1999, the Japanese government spent over one trillion US dollars on economic packages designed to stimulate the economy. Much of this money earmarked for public works projects went to construction companies, who in turn provided reliable support for the LDP during elections (Bowen 2003: 91).

In this way, the ruling party is able to affect dominant-power politics by allowing the formation of patron–client networks, which inevitably require control over the institutions allocating resources, namely government.

Ruling parties can also solve the problem of leadership succession. This problem is often exacerbated in more closed regimes where power struggles for leadership positions may cause a split among ruling elites (Way 2006), but also in democratic regimes, where the party may provide the mechanism for balancing the interests of internal party factions (the Japanese LDP is a good example). Ruling parties often supply a leadership cohort and regulate competition for the highest executive office, whether president, prime minister or party leader. To this end, the Mexican PRI stands out as a clear, if somewhat exaggerated, example of a hegemonic party autocracy in which the party nominated presidential candidates, but with the president only standing for a six-year period before stepping down. This non-consecutive re-election had the benefit of providing access to the highest executive office for a large number of party politicians, keeping disappointed or unsuccessful candidates hopeful (Magaloni 2006: 17).[6] In general, ruling parties contribute to dominant-power politics by controlling access to power positions and using institutionalised party procedures to limit power struggles that may otherwise split elite support for the regime.

A related aspect of this governing role is what may be termed 'command and control', in which the ruling party oversees and monitors the implementation of polices. The CPSU in the Soviet Union is, again, another extreme yet illustrative example of the way that ruling parties are often active in important or strategic sectors of the state, including the economy. Factories and state enterprises were subject to a complex system of supervision, including the discipline of the plan, and a highly centralised ministerial chain of command and police control, but also party control (Fainsod 1963: 506–7). The party assisted factories and enterprises in reaching output targets, but also signalled deficiencies and abuses in management, acting as an extra and trusted layer of administration. Other ruling parties have also extended their governing role to include overseeing economic development, such as the Taiwanese KMT, which controlled many strategic sectors of the state such as business, state employees, labour and farmers (Chu 2001: 292). Some ruling parties have taken this one step further, designating society itself and not just the economy or political institutions as a strategic sector to oversee and govern. In Kenya in the late 1980s it was reported that the ruling Kenyan African National Union (KANU) acquired policing functions. In 1987, KANU announced that the police would receive assistance from party members, while President Daniel Moi's New Year's Eve speech in 1989 empowered the party to monitor public places, such as bars, hotels and restaurants, and to identify regime opponents (Widner 1992: 170).

In this sense, the party's governing role may also include coercion. Coercion may function on a legal level, for example the ruling party may pass legislation that is so broad that it empowers security forces to clamp down on opposition groups deemed in violation. For example, it has been noted that UMNO's hegemony is maintained, in part, through the use of coercion, including arbitrary arrests and detention under the 'fearsome' 1960 Internal Security Act that the party helped pass (George 2007: 894). However, coercion may also work through the party itself. Youth groups and party militants may be used to harass opposition and use violence and terror to perpetuate the ruling party's dominance. In the example of the aforementioned KANU, youth wings were created by the party to patrol the country and galvanise regime support, as well as monitor dissent (Widner 1992: 132).

The implication is that ruling parties can substitute for other state institutions when need be (Way 2006), but in certain situations the party can also contribute to dominant-power politics by taking the authority of the regime to the streets. Aside from using supporters for the purposes of coercion, sometimes a spontaneous counter-mobilisation can prove decisive in regime outcomes (Geddes 2006: 11–12), especially a show of strength that counters opposition mobilisation and/or anti-regime demonstrations, to underline who is in control.

The integrating role of ruling parties – elites and society

To build and maintain any regime there must be a level of societal and elite integration, and this is just as true for democracies as for authoritarian regimes. This is summed up by the truism that the 'most important and enduring forms of collective political action in the industrial capitalist democracies are electorally competing parties' (Skocpol 1985: 24). Political parties in non-democracies, as in democracies, are forums for socialising elites, building cohesion and a sense of common purpose and an acceptance of the 'rules of the game', although not necessarily a democratic game. The party organisation is also a means for disciplining elites and providing training and other forms of indoctrination. Although the possibilities of mass socialisation and integration may be limited in authoritarian systems, in which society is often demobilised (Linz 1970), the striking examples of totalitarian systems clearly show the full integrative potential of political parties (Huntington 1968: 400; Linz 2000: 91).

In terms of elite dynamics, there are a number of challenges that incumbents face in maintaining and concentrating political control, which, in turn, directly relates to the persistence of dominant-power politics. Leaders must be able to control allies, prevent disunity and deter power challenges from other elites. As Linz notes, limited pluralism makes the authoritarian elite less homogeneous than in the totalitarian system (Linz 1970: 271), so this increases the need for integration in hybrid regimes where opposition are allowed to exist and compete in elections. Ruling parties provide the means to absorb opposition and counter elites and assimilate them into the party. Ruling parties may be ideologically strong and so offer collective benefits to attract supporters, or they many be non-ideological, offering only selective benefits of greater career opportunities, business opportunities, immunity from prosecution or any number of similar advantages. Overall, the cadre dynamics of a ruling party have been compared with the incentive structure of a 'stag hunt' game, 'in which all players are better off if they fill in their share of the circle around a stag: no one ever has an incentive to do anything but cooperate' (Smith 2005: 427).

In addition, parties often have connections with other organisations within society and so have links with large numbers of citizens through various exchange relationships (Geddes 2006: 4). These exchange relations have the potential to integrate the party within the wider society, increasing their interdependence, strengthening patron–client networks and incumbent control. The existence of exchange relations is common among many parties in democracies. A good example of the integrative potential of the party in democracy can be seen with the success of Israeli parties in integrating the Arab minority, as successive parties subordinated a potential source of regime instability through a combination of 'co-optation, repression and barter' (Shalev 1990: 109).

Ruling parties with a large or mass membership may also increase the stake and vested interests of wider society in the regime's survival. Party workers often draw salaries, have preferential access to jobs in the state bureaucracy and schooling for their children, and have 'insider' opportunities to form businesses, where connections help them secure lucrative government contracts or profit from trade restrictions (Geddes 2006: 4). Although the function of mass membership party organisations in wealthier societies may be 'vestigial', in poorer, less-educated societies, grass-roots membership is an effective means for delivering concrete benefits to supporters (Levitsky 2007: 209). Some parties increase membership to the point where the party becomes a central part of society itself. The Taiwanese KMT was able to consolidate its dominance through an inclusive member recruitment policy that saw party membership peak at a sixth of the adult population (Chu 2001: 269).

The integrative role of parties can also be seen in the contribution they make to ideology and the collective benefits offered by the party, as the basis for durable support. The ruling party may succeed in embedding itself in a larger historical moment in the life of the nation, perhaps as the emancipator of the people, such as the African National Congress (ANC) in ending apartheid or a number of nationalist parties in ending colonial rule. Over time, this ideology may form a positive association in the minds of voters, connecting the ruling party with economic success, prosperity, conflict resolution, stability or any number of events that enhance the party in the eyes of the electorate.

UMNO represents a good example of the way that a ruling party can integrate through its ideology and so perpetuate dominant-power politics, in terms of both generating cohesion, but also keeping certain groups in society in check. From 1946, UMNO adopted the role of protector of the Malay people to further the party's own political interests, in effect creating a Malay tradition that did not exist before (Singh 1998). In fact, considering the dominance of UMNO and the importance of this tradition it may seem odd that Singh (1998: 243) mentions that, before 1957, it was difficult to speak of the existence of 'a Malay' let alone a 'Malay tradition'.

Aside from creating a sense of unity, incumbents have also utilised 'ideas' to counteract potential opposition. In 1999, UMNO launched a series of propaganda measures targeting writers and academics, while Malaysian state officials propagated the idea that school teachers were 'sowing hatred of the government among their students', resulting in the requirement that university freshmen from 1999 attend a series of state propaganda sessions (Heryanto 2003: 32). In a similar way, Huat (2007: 924–5) highlights how the ruling People's Action Party (PAP) in Singapore effectively created a favourable political culture, one that had resonance among ordinary Singaporeans and which contributed to dominant-power politics without the need for repression. This political culture has played on ethnicity in a multi-cultural state, but stressing PAPs ability to govern in the best interests of society as a whole, rather than the interests of any narrow group. This in turn suggests that only PAP, not the wider electorate, knows what Singapore society's best interests really are.

Ultimately, it is the synergy between electoral, governing and integrative roles that creates the integrative power of ruling parties. Elections not only provide legitimacy for incumbents, they also serve to screen supporters and opponents, providing the incentives for politically and economically active segments of the population to join the party and stay loyal (Diaz-Cayeros *et al.* 2003: 35–6). If a political or business career depends on the outcomes of elections, and if exiting the party is likely to impact negatively on an individual's career, then the ruling party acquires strong selective benefits to offer supporters and, at the same time, control them with threat of expulsion from the party. These selective benefits and sanctions are important for non-ideological ruling parties, such as the Mexican PRI, where power and money represent the real rewards for political entrepreneurs (Reding 1991: 268).

The electorally successful ruling party, with its near monopoly over the distribution of benefits, the provision of jobs and the ability to make decisions of national importance, attracts greater numbers of supporters and, in doing so, condemns opposition parties to a vicious cycle of dwindling importance (Rasmussen 1969: 409). As for the virtuous circle, it is often a case of reciprocal causation. As noted in the case of India's ruling Bharatiya Janata Party (BJP), explanations for the rise of the ruling party are found at the system level, the interaction of the party-system and electorate, but, as the party rises, it then begins to exert and influence this very same system (Seshia 1998: 1037).

Anyone familiar with Russian politics will find immediate resonance between many of these examples and the case of United Russia. The following chapters consider these aspects in more detail. The important point to make is that, by 2010, the academic literature contains a great deal of information documenting how ruling parties generate dominant-power politics, even in political systems characterised by regular competitive elections and opposition parties. The theoretical framework outlined above presents an ideal-type ruling party – one that plays strong election, governing and integrating roles – although ruling parties will inevitably perform these three roles to differing degrees. In fact, the final section of this chapter now considers this point in more detail to understand when a ruling party plays these roles to a significant degree and therefore when the mechanisms linking ruling party with dominant-power politics are activated.

Varieties of ruling party

In addition to the studies by Brooker (1995), Way (2006) and Levitsky and Way (2010) on the role of parties in non-democracies, a range of literature has appeared in recent years building on classic studies (Huntington 1968 in particular) to 'bring the party back in' and reassert the importance of parties as explanatory variables in macro-political outcomes (Alexander 2004; Angrist 2006; Brownlee 2007; Corrales 2002; Geddes 2003; Hadenius and Teorell 2007; Levitsky 2003; Waldner 1999, Way 2005, 2006). These studies and the ones referred to in the previous section serve as a primary rationale for the framework employed in this study.

However, it should be noted that another rationale for examining the electoral, governing and integrative roles of United Russia in relation to dominant-power politics is that these same broad roles can be seen with ruling parties in democratic systems too. Normative considerations of the desirability of democracy aside, every state and political system must deal with similar power-related issues, so ruling parties in modern electoral democratic as well as electoral authoritarian systems share common spheres of activity, even if the way they go about the business of electioneering, governing and integrating differs. As Juan Linz noted, the problem of maintaining control and gaining legitimacy are common to all political systems (Linz 1970: 254).

This fact is most vividly demonstrated in the case of those exceptional democracies already mentioned, in which a ruling party, operating within a context of multi-party competitive elections, nonetheless retains power over the course of several election cycles, in some cases several decades. The examples of ruling parties in post-war West Germany, Italy and Japan, especially during their respective post-authoritarian stabilisation periods in the first two decades following the end of the war, show that dominant-power politics may, in some cases, be a requisite for later democratic consolidation. In a similar way, political parties in many twentieth-century dictatorships were not unique authoritarian creations, but simply extended their earlier use in democratic systems to achieve a monopoly over politics and society, 'either in law or in effectiveness' (Brooker 1995: 7). This implies that the roles played by parties in democracies and these modernised dictatorships were essentially the same, but their scope significantly increased in the latter.

As mentioned, Russia, 2000–10, has maintained a reconfigured democratic discourse from the Yeltsin period, while the political order that is presently consolidating is doing so within the formal framework of existing democratic institutions. Elections in the post-Yeltsin period may not be free and fair, but they are nonetheless as important in Russia as in any democracy in terms of the legitimacy and power stakes involved.

Ruling parties and agency theory

Although many of the examples given in the previous section are very contemporary, this literature is not altogether new, as examples of electoral authoritarianism pre-date the recent post-Cold War experience and can be seen throughout the three waves of democratic transition. Both the Mexican regime with the ruling PRI (1946–88) and the Argentine regime with the ruling Peron Party (1945–56) operated in systems that combined elements of democracy with authoritarianism. These regimes, in much the same way as the current Russian regime, represented a challenge to observers in determining if the democratic glass was half empty or half full. For example, before the 1950s, Mexico was considered authoritarian by the relatively few people familiar with Mexican politics, while in the 1950s and 1960s the country attracted a range of regime classifications, including authoritarian, democratic, 'imperfectly democratic' and 'in transition to democracy' (Purcell 1973: 29).

In fact, the Mexican case has relevance for this study for a number of reasons, among them the belief that Mexico and the PRI represent a good comparator for understanding Russia and United Russia (Gel'man 2005, 2006, 2008; Gvosdev 2002; Riggs and Schraeder 2005). Mexico and the PRI are also useful for answering a remaining, but crucial, question for the framework presented thus far: is it possible that a ruling party plays election, governing and integrating roles, but that the mechanisms described do not work as theorised, that the party relationship with dominant-power politics is different? It would be wrong to assume that every ruling party has the same innate capacity to affect macro-political outcomes, so there are clearly deeper processes at work. The aim of this final section is to reconsider these mechanisms from the perspective of when and under what conditions they are activated.

Although many existing studies identify ruling parties as important factors in macro-political outcomes, there is a (often implicit) suggestion that certain kinds of ruling party are more consequential than others. For example, Brownlee notes that the longevity of authoritarian regimes is affected not only by the presence or absence of a ruling party, but also by the nature of the ruling party. The ruling party that Brownlee refers to is one that controls elites and contains contestation within the private party arena (Brownlee 2007: 131), can attract and amalgamate potential opponents (p. 131), can prevent elite defection owing to its access to prestige and power (p. 154) and sets the national agenda (p. 157).

In the case of the Mexican PRI, the party appears to score highly on each of these counts. The party, which lost control of the presidency for the first time only in 2000 and which traces its roots to the early years following the 1910 revolution (Scott 1959), successfully managed elections, controlled the nomination for presidential and other crucial power positions, created and controlled patronage networks, manipulated economic conditions to its own advantage and used its central place in the history of the Mexican Revolution to cultivate sustained electoral support (Magaloni 2006; Scott 1959). The PRI was a strong ruling party across a number of indicators, one that was able to effectively control elites and set the national agenda, in line with Brownlee's elaboration mentioned above.

In this sense, the ruling status of the PRI seems to tally with existing notions of party strength found elsewhere in the literature. For example, Huntington (1968) considers a party strong if it has institutionalised support and organisational complexity, but also to the extent that activists and power seekers identify with the party (Huntington 1968: 408–12). Way (2005) measures party strength through the concepts of 'party scope' and 'party cohesion'. Scope refers to the 'size of the party's infrastructure, or the degree to which it penetrates the national territory and society', while cohesion refers to 'incumbent's ability to secure the cooperation of partisan allies within the government, in the legislature and at the local or regional level' (Way 2005: 17). Leader charisma, patronage ties and a shared history of struggle represent sources of party cohesion (Way 2005: 17–18).

There is no doubt that these notions of party strength are useful for understanding the overall contribution of parties as explanatory variables, but the emphasis on de-contextualised indicators is problematic. A party may have an extensive organisation, have a large membership and show conclusive signs of dominance over the party-system, but not necessarily have dominance over real power in the political system. A party may be labelled 'ruling' and appear to set the national agenda, but in effect be subordinated to other institutions and/or individuals beyond the control of the party. This point can be understood clearly through a commonly used concept from institutional economics – the principal–agent distinction. An agency relation involves a principal delegating authority to an agent in order for that agent to perform some kind of task (Kiser 1999: 146). Agency theory is also defined in terms of cooperative behaviour that entails an inherent conflict of goals, different attitudes towards risk and the problem of how the principal verifies what the agent is actually doing (Eisenhardt 1989b: 58). This is captured well by North's extreme example of the master–slave relationship, in which the master must invest resources to monitor the output of the slave and apply rewards and punishments in what is an implicit contract between the two (North 1990: 32).

These points, of an unequal relationship and, at times, conflicting goals, are actually very relevant for the case of United Russia and are raised at several junctures in the chapters that follow. However, the value of the principal–agent concept for this study does not lie in exchange relations, but in its ability to discriminate between types of ruling parties based on their status as either principal powers or agents of power-holders in a political system. In this sense, United Russia as a party agent corresponds to its characterisation as a 'subaltern body' (Sakwa 2010: 28), an appendage to power, rather than a central force in the political system. This relationship, in turn, serves as the important contextual factor for understanding the mechanisms detailed above, a factor strongly affecting the ruling party–dominant-power politics relationship.

Overall, agency theory and the associated problems of cooperative behaviour already form a mainstay of the overarching theory of parties in democracy, seen in terms of voters (principals) delegating their authority at election time to competing parties (agents) to represent (or not) their interests in government and policy-making (see Downs 1957 for a general elaboration on this point). By extension, the ruling party in a non-democratic context, overseeing sub-competitive elections, diverges from this democratic ideal, typically appearing as the principal power in the political system, delegating authority to its own party agents who are collectively appointed to run the country, but in the name of the party, not the people.

What is important to note is that, even in these non-democratic systems, some nominal ruling parties are, in fact, agents of independent power-holders, receiving their authority from individuals who remain beyond party control. Although a relatively under-researched area of both party politics and regime studies, such parties are often created by the military as civilian fronts for the regime (Case 1996a, 1996b: 116) or are the result of party rule degenerating into personalist rule, in which power is systematically removed from the party (Brooker 1995: 9–10).

This study shows that there is also an alternative way that the principal–agent relationship can be reversed, aside from when a military regime creates a party front or when party rule degenerates into personalist rule. This alternative is when power-holders in a civilian regime decide to create a party to serve their interests and perform certain functions, but without ceding significant, if any, executive power or authority to the party in question. Although there are studies that look at parties created by the military as civilian fronts, such as the Indonesian Sekretariat Bersama Golongan Karya or Sekber Golkar (Reeve 1985) and the Syrian Ba'ath party (Heydemann 1999), and although there are numerous examples of party rule degenerating into personalist rule, in particular in the African context, few studies have considered modern ruling parties created by a civilian executive, less the role they play.[7] This is despite the presence of several examples, notably in Peru under the rule of Alberto Fujimori and, of course, the party of power in several post-communist countries, including Russia. Although the reversal of the principal–agent relationship between party and leader has been referred to in the literature (Brooker 1995: 10), to date no study has considered this in detail.[8] The conclusions to this research return to both points.

The shifting balance of party power over time often obfuscates the principal–agent relationship, making this a difficult area to research and leaving observers unsure if leaders are or are not party agents. A good example can be seen with the Colorado Party in Paraguay, which ruled in a number of guises during its six decades of dominance, with more than a hint of changing principal–agent relations. The party first ruled as a 'civilian hegemonic party' from 1947 to 1954, then evolved into a 'military–civilian authoritarian regime' between 1954 and 1989, but following a coup in 1989 transitioned back to a 'civilian hegemonic party' until it was finally defeated in 2008 (Abente-Brun 2009: 144).

Likewise, the Mexican PRI draws an array of assessments in terms of its status as either principal or agent during its tenure as a ruling party for much of the twentieth century. The first official revolutionary party was formed in March 1929 by the military regime but in the period 1934–40 'it became clear that political power lay with the official party and not with any single leader' (Scott 1959: 108). To get an idea of the subsequent dominance that the PRI enjoyed, the following figures are useful: in 1982 the PRI controlled 91 per cent of all political posts in the country, including the presidency, 100 per cent of Senate posts, 74 per cent of deputy posts, 100 per cent of governorships, 76 per cent of state legislature posts and 97 per cent of municipal presidential posts (Casar 2002: 136).

However, the PRI continued to receive varying interpretations in the literature, with some scholars unsure if the party was really anything more than an appendage of power, an agent of individuals operating independently of the party. Without listing these different opinions in their entirety, it is useful instead to consider some of the points made in a 1969 article (Needleman and Needleman 1969: 1027) on the subject of who rules Mexico. The seemingly incompatible scholastic answers include (1) the president rules unhindered by other groups; (2) state bureaucracy rules; (3) the PRI rules; or (4) a combination of groups rule, in what the authors call the 'stalemated president' option. The authors conclude with a useful explanation for this confusion, quoting C. Wright Mills and his analysis of power distribution in the United States: 'we must expect fumbles when, without authority or official aid, we set out to investigate something which is in part organized for the purpose of causing fumbles among those who would understand it plainly'.

Context and the mechanisms of dominance

Determining if the ruling party is a principal or an agent is by no means an easy task, but the argument presented here and elaborated in the chapters that follow is that this contextual factor (where power lies) strongly affects how the mechanisms of dominance work. To paraphrase Richard Rose, the potential of parties as explanatory variables very much depends on whether they reside in the house of power as 'a master, prisoner, courtesan or eunuch' (Rose 1969: 413). As emphasised throughout this work, United Russia, as an agent of independent power-holders in and around the federal executive branch, is, in effect, the result of dominant-power politics, not its cause.

For ruling parties to generate dominant-power politics in the way discussed previously in this chapter, the party needs to be a principal power in the political system, and the only way to determine this is to examine both the origins of the ruling party and the structure of power in the given political system. One of the drawbacks of the recent move towards (re)classifying non-democratic regimes in terms of electoral authoritarianism or other primarily 'quantitative' categories based on electoral competition is that they give little information about the way that power is actually organised within the regime. Although they tell us something important about democratic development, they do not, for example, tell us whether power is organised along party, military or personalist lines, as in the widely used qualitative regime typology elaborated by Geddes (1999).

In a similar way, indicators of party dominance over the party-system, such as consecutive election majorities, do not always mean that power in the political system is organised along party lines. In fact, the treatment of party dominance in existing literature resembles the quantitative regime classification schemes just mentioned. The only consistent distinction made on the subject of party dominance is between dominance in competitive versus non-competitive party-systems (Magaloni 2006; Pempel 1990b; Thackrah 2000). The reason for this, one may surmise, is that, in an overwhelming majority of cases, parties dominant in the party-system are also ruling parties and principals in the political system as a whole.

However, existing literature, in particular the authoritative work on party dominance by Sartori (1976) and Pempel (1990b), identifies the importance of power in considerations of dominant and hegemonic party types. Although Sartori refers to electoral majorities and the number of successive election victories in his definition of 'predominant party-system' and 'hegemonic party', overall his elaboration of party dominance has a distinct power dimension too. This can be seen in the fact that Sartori identifies a specific 'ambassador-type' party as one that is clearly an agent of power, one that is external to, and uninvolved with, government (Sartori 1976: 17) and so afforded little real authority. This echoes Pempel (1990b: 3) and his observation that a party 'must enjoy a dominant bar-gaining position' to be considered dominant, meaning that the party must control government and executive power.

According to Geddes (2003), what distinguishes between real and nominal ruling parties is the extent to which they exercise some power over the leader, 'control the selection of officials, organise the distribution of benefits to support-ers and mobilise citizens to vote and show support for party leaders in other ways' (Geddes 2003: 52). This power analysis, along with an exploration of the modus operandi of the ruling party, forms the basis for understanding the principal–agent relationship in more detail and so the power of parties as explanatory variables.

Finally, it is important to note that principal–agent relations may change over time, as is seen in the case of the PRI and Colorado Party, and that, in any given political system, context is crucial for understanding the explanatory potential of the ruling party. Principal–agent relations are not always singular, so the possibil-ity of multiple principals and agents must also be entertained (Kiser 1999: 156). At the same time there is also a question of how much authority the principal actually gives to the party agent. Gandhi and Przeworksi (2006) highlight the way that authoritarian regimes use democratic institutions, notably elections, in an instrumental way to perpetuate their power. From this perspective, it is entirely possible that the ruing party has very little authority and is simply used as a tool by independent or non-party power-holders. In situations such as this, when the party-power, principal–agent relationship is significantly reversed, labels such as ruling, dominant or hegemonic party really do carry little meaning.

The following chapters contribute to our understanding of a particular kind of ruling party that has thus far escaped detailed analysis. However, there is another important contribution that this study makes to our understanding of parties in relation to macro-political outcomes. Although the rise of dominant-power politics in the post-Yeltsin period and the growing dominance of United Russia co-occur, the relationship between ruling party and dominant-power politics is very different from the one theorised. The strength of United Russia actually comes from the strength of powerful, independent or non-partisan individuals, reversing the relationship between party and outcomes theorised. This point forms the substantive argument of this research as a whole, one that is demonstrated in each of the following chapters. United Russia in the post-Yeltsin period, 2000–10, is not a true ruling party or a truly dominant party. United Russia is simply a preferred party agent of power-holders, a reflection of the intentions and ability of powerful individuals to project their power onto the party.

Concluding remarks

Overall, the period 2000–10 is characterised by political consolidation and the advent of dominant-power politics, in which one group monopolises power, despite the presence of opposition and elections. Dominant-power politics in the post-Yeltsin period resembles electoral authoritarianism rather than electoral democracy, seen in a number of political processes, but in particular the electoral process. Instead of looking at the party's contribution to democratic development, this study looks at the party's electioneering role (Chapter 3), governing role (Chapter 4) and integrating role (Chapter 5) in line with the framework presented.

Ruling parties are far from homogeneous, exhibiting varying relationships with power. Understanding the potential of parties as explanatory variables and the mechanisms that link ruling parties with dominant-power politics requires an appreciation of context, and this forms the subject matter of the next chapter. By engaging with the origins of United Russia and the party of power phenomenon we see how and when party agents, with little authority, appeared in post-Soviet Russia. Although power changes over time, party origins provide an important indication of whether the organisation is created to hold power or simply as a front and instrument to further power.

2   Parties of power
The origins of United Russia

In comparison with United Russia, Our Home is Russia was a kindergarten. So, Russia's Choice was a day nursery. Our Home is Russia was a kindergarten, Unity was a school and United Russia is an institute.

(Sergei Popov 2007)

A party of power is a party created and built by the Kremlin, a party that completely supports the Kremlin with the aim of transferring the Kremlin's power to the State Duma.

(Igor Bunin 2007)

This chapter makes three contributions to our understanding of United Russia in the post-Yeltsin period. First, it clarifies the opaque party of power term by redefining it as the 'party agent' ruling party discussed in the previous chapter. Second, it traces the origins of United Russia to show how the historical circumstances of the late Soviet and early post-Soviet periods led to the creation not only of party agents, but also of party agents with little independent power, strongly reliant on power-holders in and around the federal executive branch. Third, it answers the secondary question dealt with in this study – how United Russia appeared as a large, all-national, electorally successful party when a great deal of literature from the 1990s identified a number of serious hindrances to party-building in post-Soviet Russia.

The first part of this chapter begins by analysing the party of power term, highlighting some of its ambiguities, but also the way its defining features actually support the party-agent party type. Part two extends this discussion to examine the first attempts at creating party agents in the late Soviet and early post-Soviet periods, leading to the appearance of the Russia's Choice bloc in the founding State Duma election of 1993. This section then looks at the problems of creating reliable pro-government support in the First Duma Convocation (1994–5) as part of the background to the critical juncture of the 1995/6 election cycle when the party of power entered its current path of development.

The third part of this chapter examines 1995/6 in more detail as the moment when the party of power assumed the characteristics that it has today. This juncture had the effect of 'endogenising' party of power development; that is, making its success and failure more reliant upon the endogenous elite variable, as opposed to exogenous societal or 'environmental' factors. The final part of this chapter considers the 1999/2000 election cycle as the start point of the post-Yeltsin period and the rise of Vladimir Putin. The 1999 State Duma election also provides further confirmation of the effect of the elite variable on party of power development. In the case of the 1999 election, we see how the intentions and ability of power-holders to make the party of power project work are essential for understanding the success of Unity, but also for explaining the subsequent development of United Russia.

The Russian party of power

As this chapter is primarily interested in accounting for the origins of United Russia and highlighting the development of party agents in Russia, it is logical to begin this account with the party type that has been common currency in the entire post-Soviet period – the party of power. A party of power, according to the definition offered in the introduction chapter, is a political organisation created to support executive power and so already there is a strong suggestion that parties of power and party agents are one of a kind. This definition, in line with other definitions, also identifies certain implicit and explicit characteristics of this party type, such as their relationship with power, their functional logic and their top-down as opposed to bottom-up nature (Colton and McFaul 2003: 48; Golosov and Likhtenshtein 2001: 7; Smyth 2002: 556). Existing definitions also imply their reliance on the executive branch (Bacon 2004: 42) as well as on the resources of the state (Gel'man 2005: 5), and all these features are elaborated in more detail in the material that follows.[1] However, any deeper investigation of the party of power phenomenon encounters a number of conceptual issues that complicate the task of specifying the antecedents of United Russia with any degree of certainty. The following paragraphs explain.

Case selection: identifying parties of power

A typical functional definition of a party of power, one distilled from the literature and expressed in the most parsimonious terms, is of an elite group that supports executive power. The problem with this definition as well as other more complex definitions is that there is always a broad or narrow interpretation or application that is possible. This is relevant as it represents an immediate hurdle to identifying the origins of United Russia within the larger party of power phenomenon. Broad interpretations of the party of power potentially include many, even most, Russian parties, while narrow interpretations, as shown, are based on some rather arbitrary considerations. This makes both interpretations problematic, so in many ways the party of power, such as United Russia, is a puzzle in its own right – what exactly is a party of power, its characteristics and reason for appearance in the first place?

Of the two interpretations mentioned, it is the broader interpretation that causes the most obvious problem. Broad interpretations of the party of power make consistent case selection almost impossible, as they typically include non-party actors, what Colton and McFaul (2003: 48) refer to as members of 'the establishment'. Party of power in its broadest sense describes all elites, regardless of their political orientation, who are in some way 'part of the structure of power' (Henkin 1996). As shown in the next chapter, the growing control of the executive branch over the party-system in the post-Yeltsin period means that many parties are currently part of the structure of power, and so in theory all Russian parties may be considered parties of power.

In a similar way, the alternative narrow interpretation, like the broad interpretation, fails to resolve the problem of case selection and in many ways creates additional problems. Narrow interpretations usually focus on institutionalised elite groupings (Turovskii 2006: 153), typically political parties or movements, identifiable by their closeness to power (Colton and McFaul 2000: 202) or by their status as the 'designated winner' in elections (Oversloot and Verheul 2006: 394). In this respect, the party of power is associated with Russia's Choice at the time of the 1993 State Duma election, Our Home is Russia in 1995, Unity in 1999 and United Russia in 2003 and 2007.

One of the issues with narrow interpretations is that case selection based on closeness to power remains arbitrary, while case selection based on election results to determine the designated winner is too retrospective. Often, outcomes are poor indicators of intentions, so identifying parties of power through electoral performance, despite the convenience of this method, is unsatisfactory. Overall, narrow interpretations leave the party of power status of many parties unclear. For example, Sergei Shakhrai's Party of Russian Unity and Accord (PRES) actually had four government ministers in its ranks at the time of the 1993 State Duma election, having a genuine claim of 'closeness to power' to rival that of Russia's Choice.[2] The Ivan Rybkin Bloc, created to compete in the 1995 State Duma election, was also close to power and was initially created as a designated winner, even though it was quickly sidelined in favour of Our Home is Russia as the election campaign progressed. Although the 1995 election results show the Ivan Rybkin Bloc to be anything but a designated winner (the party received just over 1 per cent of the party-list vote), it nonetheless fits the broad definition of a party created to support executive power.

Other grey areas include parliamentary deputy groups that formed mostly from independent deputies elected in Single-Mandate District (SMD) races to support government policies as unofficial parties of power. Examples include New Regional Politics, Regions of Russia and People's Deputies groups. The party Motherland (Rodina), created to take a part of the CPRF vote in the 2003 State Duma election, retains an uncertain party of power status, as does A Just Russia, which essentially operated as one half of a tandem tactic alongside United Russia from late 2006. Although opposition parties in name, both the LDPR and the now defunct Union of Right Forces at various times enjoyed close relations with key figures in and around the federal executive branch. At the same time, the ambiguity over what constitutes a party of power led some observers at the time of the 1999 State Duma election to label Fatherland All-Russia (FAR) as a claimant for the role of party of power (Gel'man 2005: 4), but also a 'shadow party of power', 'alternative party of power', 'party of future power' and 'party of power in waiting' (Colton and McFaul 2003: 79).

One explanation for this ambiguity is that existing party of power definitions are too parsimonious, permitting both broad and narrow interpretations, so requiring more detail to better clarify the term.[3] The position taken here is that it is the 'party of power' term itself, rather than existing definitions, that represents the source of the problem. 'Party of power' was never intended to be an analytical term, but from its inception in Russian media circles in the early 1990s has always been a political term; one that captures a certain abstract reality, rather than a proposition that is logical on the basis of the meaning conveyed. This fact has important implications for this research and other research on this subject; namely that 'party of power' can only ever be understood in a general and imprecise way (as the term is used in the following chapters).

To illustrate this point it is worth considering 'party of power' in analytical terms. In the first instance, and even using a narrow interpretation of only four parties (Russia's Choice, Our Home is Russia, Unity and United Russia), parties of power in the Russian Federation have rarely been parties. Although definitions of political parties are keenly contested (J. White 2006), many so-called parties of power throughout the post-Soviet period were not parties at all. Our Home is Russia, in the words of former leader, Vladimir Ryzhkov, was a movement – an association without membership (Ryzhkov 2007). If we consider other examples, then the same absence of membership meant that Unity at the time of the 1999 State Duma election was also a movement, while Russia's Choice in 1993 was officially a bloc comprising several organisations with various levels of membership.[4]

In the post-Yeltsin period, this problem is by no means resolved with the advent of United Russia and its relatively large membership. Its particular origins and development make its status as a party open to question. The popular definition of political parties proposed by Downs (1957: 24–5), for example, suggests that parties are 'coalitions of men seeking to control the governing apparatus by legal means'. As will become clear in the following chapters, United Russia at best exerts control over legislative bodies, such as the State Duma and Federation Council, but, to date, not the 'governing apparatus' of federal government and the presidential administration.

A second source of misunderstanding that arises from treating 'party of power' as an analytical term follows closely this last point, and reiterates the substantive argument of this research as a whole. Of greater importance than their ambiguous 'party' status is the fact that parties of power actually possess little power. Therefore, the puzzle of the party of power is seen in the possibility of claiming that all or most Russian parties are parties of power, while at the same time finding grounds to argue that a true party of power has yet to appear in post-Soviet Russia. Executive power in the Russian Federation is still not organised along party lines and this point is crucial for understanding United Russia and dominant-power politics in the post-Yeltsin period, 2000–10.[5]

The party of power as a party agent

This study takes a different approach to the party of power, one that acknowledges both broad and narrow interpretations. From one perspective, party of power is understood as a general political term that includes many unspecified and unspecifiable political groupings which in some way support executive power. As a political term, party of power applies to a number of parties and movements at the federal level, but also at the regional level in the form of regional parties of power before the centralisation of the post-Yeltsin period – a subject discussed further in Chapter 5.

While recognising its inevitably broad interpretation, it is also possible to refine the party of power term to give it greater analytical purchase and to appreciate United Russia as a party agent. The antecedents of United Russia can be traced to a point in time rather than a particular party, specifically to the 1995/6 election cycle, when Our Home is Russia and the Ivan Rybkin Bloc were created to compete in the 1995 State Duma election and when Boris Yeltsin overcame the challenge of CPRF leader Gennadii Zyuganov in the two-round presidential election of 1996. From this moment onwards, the party of power acquired its distinguishing features and embarked on a new path of development, to have implications for United Russia in the post-Yeltsin period.

To understand this point in more detail it is useful to engage with two key circumstances of party creation – the party relationship with power and the party relationship with other institutions/organisations. Shefter (1994) distinguishes between two kinds of origins or circumstances at the moment of party creation. On the one hand, there are internally mobilised parties or parties created by politicians to defend their interests, but from leading power positions within the political system. Conversely, the second circumstance sees the creation of an externally mobilised party; created by politicians who seek to 'bludgeon their way into the political system by mobilising and organising a mass constituency' (Shefter 1994: 5). In a similar way, Panebianco talks of a party's 'genetic model' or the circumstances surrounding party creation, identifying the presence or absence of an external sponsor institution as a crucial influence on subsequent party development (Panebianco 1988: 50).

In short, party origins may be said to differ to the degree that parties are created as internal parties by power-holders, as opposed to external parties by power-challengers. Party origins also differ to the degree that parties are created reliant upon external sponsor institutions/organisations or are independent parties in their own right. Both these ideas are very simple, but very effective for the purposes of understanding parties of power, party agents and, indeed, United Russia.

In terms of the circumstances surrounding party of power creation, existing definitions provide important clues. As mentioned earlier, although there have been many purported examples of parties of power in the post-Soviet period, they all share certain recurring features, notably their relationship with the executive branch. This suggests that parties of power are internally mobilised, created by power-holders to defend their interests. The second dependency circumstance is less obvious, but forms the basis of the argument in this chapter and this research as a whole. Although some commentators have suggested that parties of power are far from independent (Oversloot and Verheul 2000: 134), this research argues that, from the 1995/6 election cycle, the party of power became completely dependent on the executive branch as a sponsor institution.

In comparative terms, the top-down nature of the party of power as a creation of power-holders corresponds with existing accounts of party-building found elsewhere in the literature. Duverger, for example, identifies the rise of modern political parties with the creation of parliamentary groups in eighteenth-century France and Britain that gradually developed into fully-fledged parties with the extension of suffrage in the nineteenth century (Duverger 1964: xxii–xxvi). In the Russian case, the party of power grew from the executive branch rather than parliament (Likhtenshtein 2002), although the principle of the party emanating downwards from power remains. In a similar way, dependence on an external sponsor is by no means a unique feature of the party of power. As mentioned in the previous chapter, there are examples of parties that serve as civilian fronts in military regimes or parties that are merely appendages of power in personalistic regimes, as is the case with Sartori's ambassador party party-type. In terms of the West European experience, many denominational parties were created as the political arm of the Church (Kalyvas 1996), while many socialist parties grew out of the trade union movement (Von Beyme 1985: 60).

The important point is that, before the 1995/6 election cycle, parties of power were not completely dependent on the federal executive branch, and in the case of Russia's Choice were not created solely by power-holders. It was only from 1995/6 that the party of power became a party created to represent incumbents, but with high levels of dependence on an external sponsor institution, in this case the federal executive branch consisting of government and presidential administration and those individuals in and around this institution. This position is clarified in the remainder of this chapter, by considering the dynamics and details of the party of power's historical development.

The imperative for party support in the Russian Federation

These two circumstances of party creation are useful not only for understanding the salient features of the party of power, but also for understanding its role in the post-Soviet period. The proximity of the party of power to power-holders, but also its dependency on the these power-holders in and around the federal executive branch, supports the reversed principal–agent relationship, in which the party of power operates as an agent of the federal executive branch, without real power in the political system. As noted in the previous chapter, this runs counter to the typical principal–agent relationship in which power-holders are themselves agents of the ruling party.

The fact that the party of power, and (as argued) United Russia, are not independent power-holders and that the federal executive branch is the apogee of power in the Russian political system can be understood in the context of Russia's transition from Soviet rule. This transition was set against a broader process of departification of politics and society that began in the late Soviet period, reaching a formal institutional resolution with the adoption of the Constitution of the Russian Federation in December 1993.[6] Although the origins of Russia's presidential system lay in Gorbachev's efforts to create institutional arrangements outside of the CPSU at the end of the Soviet period (Huskey 2001: 29), it was Yeltsin who finally created a constitutional basis for executive power independent of party politics. The Constitution of the Russian Federation, in line with comparative experience, very much reflected the interests of those who framed it (Geddes 1995: 239), in this case favouring president over parties and executive branch over legislative branch, as Yeltsin looked to consolidate his power in the aftermath of the constitutional crisis and president–parliament stand-off.

The Constitution of the Russian Federation also provides an institutional rationale for the party of power phenomenon. The constitution, as indicated, endows the Russian presidency with considerable paper power vis-à-vis other institutions, notably parliament.[7] However, despite the strength of the Russian presidency, in constitutional terms, both parties and parliament are still consequential for the political system and for the full realisation of presidential power too. The bottom line is that, although Russia's presidency is, in many respects, super-presidential, institutional arrangements nonetheless make parties of power highly desirable (Golosov and Likhtenshtein 2001; Likhtenshtein 2003).[8]

Russia's political system creates certain coordination problems between executive and legislative branches, heightened by its hybrid or semi-presidential nature characterised by a dual executive of president and prime minister, as in the archetypal example of the French Fifth Republic upon which Russia's system is based. Although semi-presidentialism, in theory, combines the flexibility of parliamentary systems with the authority of presidential systems (Colton and Skach 2005: 113), the dual executive of a president, but also a prime minister responsible to the legislature, creates a number of potentially conflicting power centres. In addition, and as in the case of the French Fifth Republic, the paper power of the presidential institution is realised to its fullest extent only when buttressed by a parliamentary majority (Suleiman 1994: 149). In the absence of parliamentary majorities there is a temptation for the president to bypass the legislature and rule by decree law (Stepan and Skach 1994: 129). However, reliance upon the presidential decree is not a satisfactory arrangement as in some cases it may lead to heightened conflict between executive and legislative branches (Golosov and Likhtenshtein 2001: 8–9).[9]

In the final analysis, institutions are important for understanding part of the rationale for the party of power phenomenon, but institutions do not have independent causal power. They often, but not always, exist as intervening variables that affect the calculations of elites; in the Russian case, the decision to create party support. As most presidential systems have legal provision for a separation of powers, there is often a ready-made rationale for the rationalisation of executive and legislative interests, which means that the occurrence of parties of power should be fairly common.

Although President de Gaulle created the Gaullist Party in the semi-presidential French Fifth Republic, many examples of parties of power are found in the post-Soviet space, although even here they show some variety and are by no means universal (Gel'man 2006). Belarus, for example, has a similar political system to Russia's, with strong powers afforded to the president by the constitution, but, despite the presence of dual executives and a prime minister accountable to parliament, parties of power have, to date, not emerged (Gel'man 2006: 62; Leshchenko 2008: 1430).

The mechanisms that condition the appearance of parties of power are therefore more complex than a simple institutional imperative, and this point is taken up at several junctures in the material that follows. In the Russian Federation, political and economic pluralism and elite fragmentation contributed to problematic executive–legislative relations at the federal level for much of the 1990s, reinforcing the desirability of party support for the executive branch. It is worth noting that Russia, unlike all other post-Soviet states (including Belarus), emerged from the collapse of the Soviet Union as a federation rather than a unitary state and this undoubtedly increased the overall complexity of interests vying with each other, forming part of the rationale for creating parties of power in Russia. Overall, centre–periphery relations are crucial for understanding United Russia's development, discussed later in this chapter and also in Chapters 3 and 4 and in particular Chapter 5.

The problem of creating reliable party support

The creation of parties of power in Russia is therefore strongly influenced by the prevailing institutional arrangements following the adoption of the Constitution, December 1993. However, only from the 1995/6 election cycle did they fall completely under the control of the federal executive branch. Therefore, understanding party support for the executive branch prior to the 1995/6 election cycle provides important background to United Russia.

The appearance of the first parties of power

Party of power as a political term first entered popular usage in the Russian media around the time of the 1993 State Duma election (Ryabov 1998: 81), but the party of power as a phenomenon created to support executive power traces its origins to an earlier period.[10] Aside from the incentives offered by Russia's semi-presidential system, the instigation of competitive elections also forms part of the larger, institutional rationale behind the appearance of the first parties of power. As Golosov and Likhtenshtein note, elites with no experience of participating in elections or the new institutional conditions of post-Soviet Russia nonetheless desired successful party representation in parliament (Golosov and Likhtenshtein 2001: 6). In this sense, parties of power are strategies designed to extend executive power within the institutional confines of semi-presidentialism, but also strategies to mitigate the uncertainties of newly instigated electoral competition.

It is therefore unsurprising that the first efforts to create support for executive power accompanied the advent of competitive elections in the late Soviet period and the onset of political liberalisation under Gorbachev. Until its amendment in 1990, the leading role of the CPSU was enshrined under article 6 of the Constitution of the Soviet Union, which meant that the competing tendencies of *perestroika*, of regime reformers and conservatives, were initially played out internally as rival platforms within the ranks of the CPSU. Aside from some informal or non-state groups that emerged in the early part of the reform process (Fish 1995: 33), political association was confined to the Communist Party. The advent of competitive elections to the Congress of People's Deputies (CPD) of the USSR in 1989 allowed non-party members to compete with party members for the first time.

The CPD elections in 1989 represent an early example of managed elections and so form a Soviet-era precursor to 'managed democracy' and electoral manipulation in the post-Yeltsin period. Despite the competitive nature of these CPD elections, a proportion of candidates were chosen directly from political and social organisations subordinate to the CPSU, including the Komsomol, Trade Union Federation, Academy of Sciences and the Party itself (Remington 2001: 48–50).[11] Although the introduction of these elections was a decisive turning point in the transformation of the Soviet system, CPSU members predominated in CPD elections in both 1989 and 1990 (Brown 1996: 188).

Aside from ensuring high levels of regime support by controlling candidate nomination, there is evidence to suggest that 'manufactured' opposition parties figured in these CPD elections and in the larger plans to liberalise the Soviet political system. Brown notes that, by the January 1987 Plenum of the Central Committee of the CPSU, Gorbachev was considering measures that would go beyond intra-party democracy (Brown 1996: 167). The exact nature of these measures is open to debate, but it seems likely that, in a reconfigured Soviet system, the CPSU would retain its de facto monopoly over power but with 'licensed' opposition parties permitted to compete in elections in order to provide a façade of political pluralism. The academic Evgenii Pashentsev recalls an invitation to the Institute of Public Sciences of the Central Committee of the CPSU in 1987 when this subject was discussed;

> We were told that for the purpose of democratisation, the country needed to rely on two main parties, as it is in the United States. We were also told that the game played by these two potential parties had to be convincing.
>
> (Pashentsev 2007)

Overall, the interpretation of the two-party systems of the United States and the United Kingdom as an elite strategy for managing democracy continues to find resonance among Russia's political elite. However, the first efforts at managing competitive elections by creating support parties and manufactured opposition is found in the late Soviet period, not the Yeltsin period, although some commentators find even earlier examples. Wilson, for example, describes efforts by the police to control revolutionary opposition before 1917, efforts that included the 'shadow state' sponsoring its own political parties (Wilson 2005: 3).

In the final analysis the introduction of competitive elections in the late Soviet period did not lead to the creation of a two-party political system, but efforts at creating parties to provide support for power-holders were nonetheless evident. Among the many organised interests and movements that proliferated towards the end of the Soviet period, some exhibited distinctly top-down characteristics as projects of either the CPSU or the Committee for State Security (KGB) or both. Between 1989 and 1991, several pro-regime movements emerged to counteract the democratic movements that opposed the CPSU, including Communists of Russia, formed in May 1990; *Soyuz* (Union), formed in February 1990; and the LDPR, created in the summer of 1989, and a founder of the Centrist Bloc of Political Parties and Movements (see Pribylovskii 1992).

The LDPR represents a good example of an unspecified party of power and one of the first alleged efforts to mitigate the effects of competitive elections from the late Soviet period. The origins of this misleadingly named nationalist party have been the subject of speculation for some time, leading some observers to place LDPR squarely within a much larger project to create licensed opposition for the CPSU (Wilson 2005: 22). Initially created in 1989 as the Liberal Democratic Party of Russia, the party was renamed the Liberal Democratic Party of the Soviet Union at its Founding Congress in March 1990 (Pribylovskii 1992: 44). The party supported Vladimir Zhirinovskii's candidature in the Russian Socialist Federative Soviet Republic (RSFSR) presidential election in spring 1991 and then reverted to its original name, the Liberal Democratic Party of Russia, for the 1993 State Duma election. The connection between LDPR and the Soviet authorities arises, in part, from the alleged KGB background of party leader, Vladimir Zhirinovskii. Although the original purpose of LDPR is difficult to ascertain with any degree of certainty, it has been suggested that the party was created by the CPSU and KGB to discredit and divide the emerging democratic forces (McFaul and Markov 1993: 243; Wilson 2005: 23).

Like many parties created to support power in the post-Soviet period, there is little prospect of conclusively substantiating the origins of LDPR. A senior figure in the LDPR leadership, Egor Slomatin, commented that, at the time of the party's appearance in 1989, any connections with the KGB were considered negative and so were exploited by opponents. At the same time, consultation with the local department of the KGB was an informal, but necessary, procedure for these early Soviet parties (Slomatin 2007). This may explain why Zhirinovskii went to great lengths to dispel rumours of his own KGB connections, but not the privileged access that he and his party enjoyed to top government officials and elements of the Soviet media. Pribylovskii (1992: 44) reported that LDPR's Founding Congress in March 1990 was covered by all nationwide newspapers and by the television programme *Vremya*. However, Zhirinovskii rejected claims that he had connections with the KGB. In autumn 1991, Zhirinovskii threatened to sue the author of a book that included a chapter entitled 'the KGB's disciple', while in August 1991 the KGB produced a signed document stating that Zhirinovskii had never worked for the KGB (Umland 2006: 198–9).

An equally important tendency that emerged from the late Soviet period, which was to have a large bearing on party of power development in the Russian Federation, were the democratic movements that previously formed the mainstay of the CPSU opposition and later pro-Yeltsin support in the early post-Soviet period. Rather than created in a strictly top-down fashion, these democratic movements were a mixture of CPSU counter-elites and non-party regime opponents; part of the first wave of Russian democrats. For example, the Democratic Party of Russia that appeared in May 1990 included a loose assembly of voters' associations that campaigned under this nationwide umbrella organisation, as well as individuals from within the Interregional Deputies Group that formed in the CPD (Urban and Gel'man 1997: 181). The Democratic Russia Movement that emerged in October 1990 was also an umbrella organisation linking grass-roots democratic support and Soviet counter-elites.

In the late Soviet period, it was these democratic movements (mainly Democratic Russia) that coordinated demonstrations throughout the Soviet Union, that aided the election of Yeltsin to the office of President of the RSFSR in June 1991 and which rallied millions to defeat the communist coup d'état in August 1991 (McFaul and Markov 1993: 136; Urban 1992; Urban and Gel'man 1997: 183). In June 1993, several of these democratic tendencies coalesced to form the bloc, Russia's Choice, to compete in the December 1993 parliamentary election, forming a pro-government and pro-Yeltsin party of power that combined prominent government officials and grass-roots members. Although senior government officials, such as deputy prime minister, Egor Gaidar, dominated the party (Golosov 1999a: 94), it was Democratic Russia, with its substantial links with society, that formed the bulk of Russia's Choice. As if to emphasise this point, by April 1991, Democratic Russia had built a mass base that boasted over a million members and party representation in a thousand towns and cities across Russia (Urban 1992: 191).

Russia's Choice, with its origins in the democratic movement of the late Soviet period, came to be seen as the first party of power of the post-Soviet period. At the time of the 1993 State Duma election the party was both close to power, in the sense that party leader, Egor Gaidar, held the post of deputy prime minister, but also the designated winner, polling more votes than any other pro-government party (see Table 2.1).

As previously mentioned, although Russia's Choice was the largest pro-government party, it was not the only contender for party of power status. PRES, created in the summer of 1993 by then Minister of Nationalities, Sergei Shakhrai, also enjoyed a close relationship with Yeltsin's reformist government. PRES benefited from the support of prime minister, Viktor Chernomyrdin, and other leading figures within the executive branch, including Aleksandr Shokhin, the

Table 2.1 State Duma election results, 12 December 1993

Party	Party-list vote (%)	Seats (party-list and SMD)
Russia's Choice	14.5	70
PRES	6.3	19
Yabloko	7.3	23
Democratic Party of Russia	5.1	15
Women of Russia	7.6	23
Agrarian Party of Russia	7.4	33
CPRF	11.6	48
LDPR	21.4	64

Source: http://www.russiavotes.org.

deputy prime minister responsible for foreign economic affairs, as well as Sergei Stankevich, an adviser to Boris Yeltsin (Treisman 1998a: 146). Although a pro-government and pro-reform party, PRES was more conservative than Russia's Choice and in ideological terms provided the first example of what was then termed 'new conservatism' (Ermolin 1993), a mixture of liberalism and conservatism that was to surface again in United Russia's ideology in the post-Yeltsin period (see Chapter 5).

The Women of Russia movement also gained seats in the December 1993 election and, like many of the unspecified parties of power of the post-Soviet period, was an organisation allegedly supported by the presidential administration. Women of Russia included the Association of Female Entrepreneurs, the Union of Women in Russia and the Union of Women in the Navy. The movement was led by Ekaterina Likhova, a presidential adviser on matters pertaining to women (Pel'ts 1993). Golosov (1999b: 105) notes that many observers speculated that Women of Russia was created specifically to capture a proportion of the female electorate that may have otherwise voted for the communist opposition.

The demise of the democratic movement as a viable party of power

The first parties of power in the Russian Federation were therefore a mixture of structures, but the appearance of Russia's Choice signified a party of power of a different kind; one with links to society – party origins that were both bottom-up and top-down. Consequently, Russia's Choice, unlike many of the parties competing in the 1993 elections, was, by Russian standards, a genuine proto-party, one with real potential to develop further and become a bulwark for the fledgling, post-Soviet democracy. As such, an important question for this study is why the Yeltsin administration decided to create new parties of power during the 1995/6 election cycle.

Despite the resource advantage that Russia's Choice enjoyed over its competitors, its vote share in the 1993 State Duma election was less than anticipated by those in and around the federal executive branch. Consequently, the lack of popular support for Russia's Choice is one obvious explanation for the party's relatively poor election results in December 1993 and for the creation of Our Home is Russia and the Ivan Rybkin Bloc as alternatives in 1995. This explanation, of low levels of popular support, is one that finds resonance in existing literature on party of power development in the Yeltsin period (McFaul 2001: 1162–3; Sakwa 2005: 385; Smyth 2002: 557). However, the effects of the exogenous variable of popular support should not be overstated and this point is reiterated in the next part of this chapter. To understand the fall of Russia's Choice as the party of power and the rise of the top-down, strictly subordinate parties of power from 1995 onwards, it is necessary to consider the elite variable in more detail and some of the problems faced by power-holders in the late Soviet and early post-Soviet periods in their efforts to secure reliable support in the national legislature.

In the late Soviet period, the problem of creating reliable parliamentary support for government arose almost as soon as competitive elections and pluralism were introduced. Although efforts to manage the first competitive elections to the CPD resulted in the predominance of CPSU members, many were supporters of radical political transformation that threatened to undermine the Soviet regime. The Interregional Deputies Group within the CPD, for example, contained a number of radicals such as Andrei Sakharov, Yurii Afanas'ev and Boris Yeltsin. In fact, Sakharov and Afanas'ev later proclaimed the Interregional Deputies Group to be in opposition to the CPSU (Pribylovskii 1992: 38).

In the First State Duma Convocation of the newly independent Russian Federation (1994–5), efforts to create support for federal government among deputies elected from SMDs also showed similar problems of reliability. One such parliamentary group, New Regional Politics, was created in January 1994 from SMD deputies to represent the interests of the government, primarily the Ministry of Fuel and Energy, but also other powerful individuals in the energy sector. This parliamentary group was later joined by other deputies who saw the potential of this branch of industry and, by April 1994, controlled 66 of 450 seats in the State Duma (Remington and Smith 1995: 467). The group was led by Vladimir Medvedev, the head of the Union of Gas and Oil Producers – a figure with strong connections with the Ministry for Oil and Gas. Andrei Loginov, who worked in the Ministry for Oil and Gas, 1993–4, and who was partly responsible for establishing New Regional Politics, recalls how in the autumn of 1994 he lost control over the group. 'For me it was absolutely unexpected. I looked at the voting result on an issue that was of principal importance for the government and saw . . . my New Regional Politics had split into two parts' (Loginov 2007).

In a period of regime uncertainty, as was the case for much of the 1990s in Russia, creating reliable support within the State Duma was difficult, especially with the constellation of economic interests that were vying for influence in the new political system. However, the problem extended to political interests too, evident in the growing realisation on the part of the Yeltsin administration that the democratic movement had outlived its usefulness, and that it was increasingly unreliable and outspoken. Overall, reliability was a big problem in the First State Duma Convocation, which proved to be particularly fluid with deputies frequently moving from one faction to another. However, even when deputies belonged to a faction, there was no guarantee that they would vote in a disciplined manner. The overall result was that parliament, in this early stage of its existence, was able to pass only a small amount of legislation (Belin and Orttung 1997: 21).

The First Duma Convocation, and the democracy movement as a whole, contained an eclectic mix of personalities, some of whom had strong opinions on how the Yeltsin administration should govern. As the amorphous democratic movement began to form towards the end of the Soviet period, separate ideological tendencies became evident and the movement quickly began to split. The differences between key individuals within the anti-CPSU, pro-democracy camp could be seen in the failure of the various democratic elements to conduct a unified campaign for the State Duma election of 1993. Although there was an official declaration of cooperation just prior to the election between Russia's Choice, Yabloko, PRES and the Russian Movement for Democratic Reform, led by St Petersburg mayor, Anatolii Sobchak (Pel'ts 1993), in reality the pro-democracy parties were divided (a recurring theme in post-Soviet Russia). Ultimately, Sergei Shakhrai, Nikolai Travkin and Grigorii Yavlinskii all had their own particular views on democratic development in Russia, but, more importantly, were all keen to lead their own parties. This was also true of Egor Gaidar who created Russia's Democratic Choice in April 1994 as the successor to the Russia's Choice bloc.

Although the strength of the democratic movement, at least initially, was its ability to unite elites and ordinary Russians, this strength also created friction. Michael McFaul recalls the early Democratic Russia movement as 'top-heavy with egotistical leaders unaccustomed to being one among equals' (McFaul and Markov 1993: 137). In the late Soviet period, this volatile mix of personalities was maintained only by a strong desire to oppose the Soviet system, but with this aim achieved the movement lost its strategic imperative. Differences appeared between those newer elites who rose to prominence in the late Soviet period and the older establishment elites who were busy reconnecting their career paths in the emerging post-Soviet reality. One prominent figure from the early democratic movement, Lev Ponomarev, recalls a conversation with Egor Gaidar in the aftermath of the 1993 State Duma election: 'Gaidar told me the following: "the Democratic Russia movement has had its day. There are too many weird people in the Democratic Russia movement – *demshiza* (democratic schizophrenics)" ' (Ponomarev 2007).

Some former Soviet-era elites who found themselves within the broad democratic movement were adept at playing power games to secure their own career advancement and so were unwilling to subordinate their interests to what they considered 'peripheral elites' and a wider party rank and file. In the final analysis this unwillingness to subordinate periperal elites was also true for the government and the Yeltsin administration in general. The start of the First Chechen War in December 1994 proved to be decisive in formally breaking the link between Yeltsin and the democratic movement. Commentators began to see a discernible split emerging between Yeltsin and Russia's Choice immediately after the arrival of Russian troops in the breakaway Chechen Republic in December 1994

(Chugaev 1995), although it would be wrong to say that Russia's Choice was completely opposed to war in Chechnya. Party leader, Egor Gaidar, admitted that he never had any ideological opposition to the Chechen War, just a 'deep felt unease' that, with the state of public opinion and the condition of the Russian armed forces, his own opposition to the conflict was inevitable (Gaidar 2007). Unable to support hostilities in Chechnya, the renamed Russia's Choice (Russia's Democratic Choice) effectively moved into opposition.

Although public sentiment towards both the Yeltsin regime and the democratic movement had hardened by the 1995/6 election cycle, it was war in Chechnya that provided the catalyst for the executive branch to create its own party support, independent of the democratic movement and of any societal interests. In the immediate aftermath of the split with Russia's Democratic Choice and the defection of part of the New Regional Policy parliamentary group, the executive branch drafted a new two-stage plan designed to shore up short-term Duma support and to ensure future support in the Second State Duma Convocation (1996–9). The first stage of the plan saw the formation of two new parliamentary groups, one subordinated to the presidential administration, called 'Stability', the other subordinated to the White House, called 'Russia'. Both were created to stand against the Communists and the New Regional Politics deputies who had slipped control (Loginov 2007). The second stage of the plan began in April 1995 when discussions were held between a number of interested groups in and around the federal executive branch on the subject of creating two new centrist parties to compete in the December 1995 parliamentary election. The result was the creation of Our Home is Russia and the Ivan Rybkin Bloc.

The significance of the 1995/6 election cycle

The strategy of creating two parties of power for the December 1995 parliamentary election revisited another, older theme from the late Soviet period already mentioned. The notion of creating a two-party system as the basis of executive branch and regime stability, similar to the American or British examples, is one of several ideas that repeatedly surface in post-Soviet Russian politics. Our Home is Russia and the Ivan Rybkin Bloc of 1995 were to find resonance again in the 2007 State Duma election, with the creation of A Just Russia to compete alongside United Russia. This tandem tactic, as discussed in the next chapter, was essentially the same left-of-centre/right-of-centre combination deployed in 1995, although under somewhat different circumstances and with qualitatively different results. In 1995, the Ivan Rybkin Bloc failed to generate any enthusiasm among the electorate, but more importantly failed to attract support from within the elite – further evidence of the importance of the elite variable on party of power development. Vyacheslav Nikonov, who was closely involved with both of the party of power projects in 1995, recalls the problem:

It was difficult for the bureaucracy to understand why they should support him [Ivan Rybkin] when the Prime Minister [at that time] was Chernomyrdin;

when he was the party of power number one, and Rybkin was just a small substitute. As a result, Our Home is Russia became the main support for the government in the Second Duma Convocation.

(Nikonov 2007)

The appearance of Our Home is Russia and the importance of the elite variable

With the poor showing of the Ivan Rybkin Bloc, Our Home is Russia became the mainstay of government support in the Second Duma Convocation. The movement consisted of a number of influential figures from regional administrations as well as deputies from the previous Duma Convocation, including some elected on the ticket of Russia's Choice and PRES, among others (Kudinov and Shipilov 1997: 13). Our Home is Russia also incorporated those reliable deputies who previously belonged to the troublesome New Regional Policy deputy group. Like Russia's Choice in 1993 and Unity in 1999 (dealt with in the final part of this chapter), Our Home is Russia was created only months before the State Duma election, holding its Founding Congress on 12 May 1995. The congress unanimously elected prime minister, Viktor Chernomyrdin, as party leader and ten days later, on 22 May 1995, the all-Russian political movement was registered with the Ministry of Justice. Table 2.2 summarises the December 1995 State Duma election results.

Both the short time frame available to form Our Home is Russia and the break with Russia's Democratic Choice over the ongoing Chechnya conflict forced the creators of the new party of power to draw heavily on administrative resources to party-build as well as the financial resources of businessman and shadow politician Boris Berezovskii (Dorofeev 1997). That is not to say that Our Home is Russia had no contacts with non-governmental organisations in the weak but

Table 2.2 State Duma election results, 17 December 1995

Party	Party-list vote (%)	Seats (party-list and SMD)
CPRF	22.3	157
LDPR	11.2	51
Our Home is Russia	10.1	55
Yabloko	6.9	45
Women of Russia	4.6	3
Communists of the USSR	4.5	1
Russia's Democratic Choice	3.9	9
Ivan Rybkin Bloc	1.1	3
Independent candidates	–	77

Source: http://www.russiavotes.org.

emerging Russian civil society. The party reportedly had representatives in 80 regions and links with representatives of local industry and commercial structures, as well as organisational linkage acquired from its incorporation of the Farmers Union, the Association of Russian Higher Education Establishments and the Association of Exporters, among others (Savvateeva 1995). The Afghanistan Veteran's Movement was another that had links with Our Home is Russia and it is very likely that there were significant ties with big business, notably Gazprom, in view of Chernomyrdin's previous association with the company. However, Vladimir Ryzhkov, one of the founders of Our Home is Russia and former party leader, acknowledged that, although around 30 per cent of the movement comprised 'public politicians, such as Ryzhkov himself, the remaining 70 per cent consisted of government staff (Ryzhkov 2007).

The result was a party completely reliant on the executive branch, totally submitted to the White House. The party of power, from the appearance of Our Home is Russia, became a phenomenon that was simultaneously more and less than a political party. Limited organisational reach and an absence of membership meant that, in classical terms, Our Home is Russia had a rather dubious claim as a political party, while its origins in executive power made it something much more. Our Home is Russia resembled a bureau or department of government, controlled and directed by the executive branch for the purposes of overcoming the problems of executive–legislative coordination, in line with the institutional rationale of the party of power outlined earlier. The one important difference between Our Home is Russia and other, more traditional components of state bureaucracy was the matter of elections; that Duma deputies, unlike bureaucrats, are elected, not appointed. However, with the increasing ability of incumbents to manage the outcome of elections this difference became less discernible.

The creation of Our Home is Russia marked a critical juncture in the story of the party of power; what historical institutionalists refer to as a crucial founding moment, one that was to constrain the path of its future development. The year 1995 marked the moment when the path of party of power development was re-established, independent of societal interests and the grass-roots Democratic Russia movement that formed the basis of Russia's Choice in 1993. Our Home is Russia represented a much purer, top-down party project, one much more reliant on power-holders in and around the executive branch. This critical juncture was significant for the future development of the party of power, including United Russia, and its ability to play the kind of role in the political system typically associated with ruling parties elsewhere.

The overall effect of this reconfigured path can be understood in terms of the endogenisation of party of power development. This endogenisation essentially gave an enhanced importance to the elite variable in matters pertaining to party of power appearance, disappearance, form, electoral results and role in the political system, and so signifies a reduced importance for exogenous or external stimuli, such as public opinion. The break with the democratic movement in mid-1994 and the creation of Our Home is Russia in mid-1995 gave power-holders in and around the executive branch greater autonomy from society. At the same time, it laid the foundations for the realisation of the paper power of Russia's super-presidential system by holding out the possibility that power-holders could rationalise executive–legislative relations with a party agent, removing an important check on executive power. Importantly, this rationalisation was possible without the need to subordinate to a party in a way that would constrain their behaviour.

Although external stimuli became less of a factor in party of power development from 1995/6, they were by no means redundant. The tumultuous political development under the presidency of Boris Yeltsin provided what may be termed numerous 'external shocks' (Harmel and Janda 1994: 265) that affected all political parties, including successive parties of power from Russia's Choice in 1993 to Unity in 1999. In a relatively short period between the founding elections of December 1993 and Yeltsin's resignation on New Year's Eve 1999, Russian public opinion was impacted by the disastrous First Chechen War (1994–6), default and financial crisis in August 1998, the setback surrounding the North Atlantic Treaty Organization (NATO) bombing of Serbia in March 1999, the Moscow apartment bombings and subsequent second war in Chechnya in 1999 and a more or less constant deterioration in living standards.[12] All of these external stimuli affected voter attitudes and the electoral strength of the major Russian political parties.

Likewise, this endogenisation of party of power development should not overlook the fact that the elite variable was an important influence on party of power development prior to 1995/6. If we consider again the electoral results of Russia's Choice in the 1993 State Duma election, we see a much greater influence of the elite variable than commonly acknowledged. Although the effects of economic reform or 'shock therapy' initiated in January 1992 continued to adversely affect ordinary Russians, and so voter estimations of Egor Gaidar and the pro-reform parties, the extent of the economic variable on the popularity of Russia's Choice is open to question.[13]

A number of factors contributed to the relatively poor results of Russia's Choice in 1993 (see McFaul 1998), but it is worth noting that these results were in no small way influenced by the efforts of the Yeltsin administration to ensure the adoption of the new Russian constitution in the referendum that was held concurrently with the Federal Assembly elections. The relatively poor results of Russia's Choice were not helped by widespread fraud, as Yeltsin supporters stuffed ballot boxes to ensure that voter turnout was above the 50 per cent required to ratify the new constitution, a constitution granting Yeltsin considerable powers. Although difficult to conclusively substantiate, it has been suggested that Yeltsin's supporters weakened the vote share of Russia's Choice by boosting the LDPR vote, in the mistaken belief that Zhirnovskii's party would perform poorly. Belin and Orttung cite political scientist, Aleksandr Sobyanin, who claimed that parliamentary ballots were falsified to tally with the constitution referendum, as it would look suspicious if more people participated in the referendum than in the parliamentary election. Belin and Orttung also note that these claims can never be fully substantiated as the Central Election Commission (CEC) never published complete results and destroyed all ballots four months after the election (Belin and Orttung 1997: 11).

In a similar way, Our Home is Russia and the Ivan Rybkin Bloc were certainly not the first parties created strictly subordinate to power-holders and their interests. In many ways, both were an extension of the previous departmental principle seen in the 1993 State Duma election. In 1993, several party projects were created from government departments, notably PRES, but also the ecological movement, Cedar. The founder of PRES, Sergei Shakhrai, was simultaneously a Deputy Prime Minister and also the Minister for Nationalities. The party was able to use the bureaucrats who worked for the Committee on Nationalities for party purposes and, reportedly, bolstered party ranks by ordering the heads of regional departments and their subordinates to join (Pribylovskii 2007). The Constructive-Ecological Movement of Russia, Cedar, was another party built on the departmental principle, formed between February and May 1993 with the active involvement of functionaries from the Department for Sanitation and Epidemic Control and business interests close to the Ministry for the Environment.

In the final analysis, the 1995/6 election cycle did not mark the point when federal level power-holders or the party of power became immune from the effects of external stimuli, including voter dissatisfaction. Despite enjoying significant resource advantages over party competitors, Our Home is Russia in 1995, like Russia's Choice in 1993, failed to meet the high electoral expectations of party-builders. However, the 1995 State Duma election did provide the blueprint for achieving reliable party support without the need to make side payments to any particular constituency in society, while exonerating government ministers and the president from the constraints of party discipline. Subsequent parliamentary and presidential elections were to see a much more effective use of administrative resources to the benefit of incumbents and the party of power. As detailed in this and later chapters, incumbent strength and their ability to make the party of power a success were steadily increasing throughout the 1990s, despite the increasing physical weakness and unpopularity of Yeltsin.

The dynamics of the Russian party of power

The rise of the party of power from a relatively unsuccessful and short-lived flash party at each of the three State Duma elections in the 1990s towards the hegemonic mass-type phenomenon of United Russia in the post-Yeltsin period represents a puzzle. As discussed in the introduction, United Russia as a successful, all-national party with a large membership seems to contradict a great deal of literature amassed on Russian party politics during the 1990s that highlights the relative weakness of parties, personified by the indifference of elites and voters towards them (Bacon 1998; Enyedi 2006; Evans and Whitefield 1993; Fleron and Ahl 1998; Ishiyama and Kennedy 2001; Kopecky 1995; Lewis 2000; Toole 2003). At first glance, the success of United Russia in the post-Yeltsin period suggests that some kind of significant shift in public opinion has occurred, or that some other major event has changed the status of parties in the eyes of voters and elites.

The next chapter deals with voter attitudes within the wider context of United Russia's electoral success. Here, and in the final part of this chapter, it is possible to draw together some of the points raised so far to answer this puzzle of how United Russia managed to defy the difficult party environment and the weak development of previous parties of power to succeed in the post-Yeltsin period. To answer this question, it is necessary to look again at the critical juncture of 1995/6 and the overall development of the party of power already detailed. As mentioned, 1995/6 marked the moment when the path of development of the party of power was reconfigured, when it became an extension of the federal executive branch and when the intentions and capacity of incumbents to project their power onto party agents acquired overriding significance. As such, the clue to the appearance and success of United Russia lies in the endogenous elite variable rather than exogenous factors. The following paragraphs examine this point in more detail.

Exogenous or external stimuli on party development come in a number of forms, including social, economic and political changes that take place outside or external to the party organisation (Harmel and Janda 1994: 266). In theoretical terms, the endogenous/exogenous dichotomy is similar to the distinction between 'environmentalist' and 'purposive action' approaches to the question of party change (Muller 1997: 293–4). Party politics literature, for example, has long stressed the importance of institutions as one such external stimuli that affects the characteristics of parties and party-systems (Duverger 1964: 239–5; Epstein 1980: 31–45; Rae 1971), and this has not escaped the attention of researchers looking at parties in post-Soviet Russia (Golosov 2004; Hale 2006; Ishiyama and Kennedy 2001; Kitschelt 1995; McFaul 2001; Moser 1999; Smyth 2006).

As discussed previously, there are clear institutional incentives for the formation of parties of power, but institutional theories are limited in their ability to explain the appearance of United Russia because of an absence of significant institutional change between the Yeltsin and post-Yeltsin periods. Aside from the adoption of new laws, including party and electoral laws (see Chapter 4 and Hale 2006), the only significant constitutional change came in December 2008 when President Medvedev approved changes to the length of presidential and parliamentary terms, leaving institutional arrangements in post-Soviet Russia remarkably stable over time (Gel'man 2005: 3). Institutional arrangements may go some way to explain the appearance of the party of power (although not its non-appearance, as in Belarus) and even its continuing logic and rationale, but not its variation over time, between Yeltsin and post-Yeltsin periods.

Another external stimulus affecting party and party-system development is the aforementioned changing voter attitudes (Epstein 1980: 27–31; Harmel and Janda 1994: 266), identified most prominently in the seminal work of Lipset and Rokkan (1967), but also considered in post-Soviet Russia (Bielasiak 1997; Kitschelt 1995; Whitefield 2001, 2002; Wyman *et al.* 1995). The theory of Lipset and Rokkan (1967) was one of the first to identify the relationship between parties and societal cleavages and how changes in the structure of society directly affect party development. According to this approach, either incremental change or more sudden change in voter alignments may explain the electoral rise and fall of parties, as well as party organisational change to try to regain equilibrium with the electorate.

This idea of changing voter attitudes has a great deal of resonance in the post-Soviet period. As already mentioned, existing literature on party of power development does indeed correlate voter attitudes with the poor results of Yeltsin-era parties of power and the decision to rebrand or remake them. However, the 1995/6 election cycle saw a strong expression of voter dissatisfaction with Yeltsin's leadership, but this was not enough to stop his successful re-election. Vyacheslav Nikonov, who worked on the Yeltsin re-election campaign, puts this in perspective: 'We committed every sin in 1996 not to let the communists come back. Of course, in any free and fair election, in 1996 the communists would have prevailed' (Nikonov 2007).

The ascendance of the CPRF in the 1995 parliamentary election and the strong presidential challenge made by party leader Gennadii Zyuganov in 1996 led not to leadership change but to leadership entrenchment.[14] This is why the 1995/6 election cycle and not just the 1995 State Duma election form the critical juncture of party of power development. The 1995/6 election cycle arguably marks the point when democratic development in Russia began to backslide and when the relationship between state and society began to recede as the post-Soviet regime began to close.

This is not to dismiss the significance of voter attitudes in Russian politics. Part of the reason for the creation of Our Home is Russia in 1995 was the poor electoral returns of Russia's Choice in 1993. Similarly, and as discussed later, the creation of Unity in 1999 was in no small part due to low levels of popular support for Our Home is Russia as the existing party of power. The relationship between voters and the ruling elite simply changed, moving away from the democratic ideal that power be decided through free and fair elections by the will of the people. Instead, public opinion became part of the overall calculation on the part of power-holders, something akin to a feedback loop in institutional theory (North 1990; Pierson 1993). For example, part of the viability test that parties of power undergo prior to election periods is whether they represent a good investment for incumbents over and above the investment required for building a new party from scratch. Even with strong incumbent capacity, turning an unpopular party into a popular one may involve exorbitant costs. In the post-Soviet period, from Yeltsin to post-Yeltsin, voters are one of several factors that influence the calculations of power-holders. As the material in the previous chapter suggested, gaining stable majorities in the absence of popular support will likely involve a degree of electoral fraud that may threaten their overall legitimacy.

If external stimuli emphasise the structure of politics, such as the effects of public opinion, institutions, etc., then the elite variable emphasises human agency, the intentions of individuals in and around the executive branch and their ability to transfer their power to party agents. The elite variable includes the ability of power-holders to control regime outcomes and consolidate the nascent political order as well.

The elite variable, seen in terms of the effect of human agency, as already mentioned, has figured prominently in party of power development from 1993, and even before this period. Yeltsin's personal decision to remain above party politics has been documented as a detriment to several parties of power. Colton mentions that the party of power never received Yeltsin's endorsement, while at the same time he never prevented cabinet ministers from competing in elections under different banners (Colton 2008: 281). The former leader of Russia's Choice, Egor Gaidar, recalls his disappointment that President Yeltsin reneged on his promise to attend the party's Founding Congress in October 1993:

> We agreed that he would come to the First Congress of Russia's Choice and clearly make the point that this is his party, which at that time I think would have assured us of the elections. And then, at the last moment, I got a telephone call from the head of his secretariat, that Boris Nikolaevich had decided . . . he . . . he decided not to phone me personally, he decided it would be better if he refrained from interfering in the parliamentary elections, so he would not attend the Congress of Russia's Choice. From my point of view, this practically sealed the party's fate.
>
> (Gaidar 2007)

By the time of the 1995/6 election cycle, the fate of the party of power was tied to the intentions of key figures in and around the executive branch and their ability to project power and influence onto their preferred party agent(s). By the end of the 1990s, increasing resources and growing experience of elections meant that power-holders were better able to make party of power projects work. This point touches on post-Soviet learning (Dawisha and Deets 2006), especially elite learning through the conduct of elections.

The significance of the elite variable, 1999/2000

The physical frailty of Yeltsin and his precarious position towards the end of his second term of office deflected from the fact that, over the course of the 1990s, incumbent strength at the federal level and in the regions was increasing. This increasing incumbent strength or capacity was, by Russian standards, incremental, a long-term process that began in the late Soviet period when political competition began to make the outcome of elections less certain than before.

The fall of Our Home is Russia, the rise of the Unity project

The 1999/2000 election cycle provides a good opportunity to see the outcome of learning and increasing incumbent strength, as Unity emerged as the successor to Our Home is Russia at the very end of the Yeltsin period. The role of Unity in the outcome of the 2000 presidential election is considered in more detail in the next chapter. Here, it is worth examining the circumstances surrounding Unity in more detail.

The federal elections of the 1999/2000 election cycle resembled what Key (1955) termed 'critical elections', or ones that seemed to matter more than others, in terms of both their immediate results and their lasting effects. Central to the concept of critical elections is a discernible and durable realignment within the electorate, something that Key investigated for several US elections. In the Russian case, the 1999/2000 election cycle represented a realignment of Russian elites and the start of the process whereby central, presidential power came close to reaching its potential. Alternatively, the elections of 1999/2000 can be framed in terms of another critical juncture in the path of post-Soviet Russian politics, but unlike the juncture of 1995/6, which was significant for party of power development, the 1999/2000 election cycle was significant for the entire political system, as the start point of the Putin period.

The 1999/2000 election cycle has been well documented in existing literature (Brudny 2001; Colton and McFaul 2000, 2003; Makarkin 2000; Oates 2000; Rose and Munro 2002) so the following paragraphs consider only those aspects relevant for this inquiry. In line with the argument developed in this chapter, the success of Unity in 1999 was a discernible manifestation of the increasing ability of power-holders to affect larger, macro-political outcomes.

The 1999/2000 election cycle, but also the entire period from the 1996 presidential elections, provides further evidence of the significance of the elite variable in party of power development. The emergence of Unity in late 1999 was predated by the fall of Our Home is Russia as the de facto party of power of the Second Duma Convocation; a fall directly related to elite manoeuvring in and around the federal executive branch. After Yeltsin's close-run re-election as president in July 1996, the attention of many highly placed figures began to focus on the 2000 presidential election as the moment when Yeltsin was constitutionally obliged to step down. As a result, and in no small way related to the physical weakness of Yeltsin, certain individuals began to position themselves ready for this opportunity.

It has been noted that Yeltsin's own style of leadership involved a system of checks and balances designed to keep competing elements under control (Shevtsova 2001: 82). Former Yabloko leader, Grigorii Yavlinskii, described Yeltsin's court as 'byzantine', with members constantly engaged in palace intrigues (White 2000: 93). Viktor Chernomyrdin, as the relatively long-standing prime minister, was no stranger to palace intrigue and was seen by many as Yeltsin's logical successor. In fact, in early 1996, it was rumoured that Chernomyrdin was considering running for the presidency alongside Yeltsin in an effort to thwart the communists, with the possibility that during the campaign one would withdraw their candidature in support of the other (Kononenko 1996). If this is true, it provides further evidence of a longer-term interest in tandem tactics among the Russian elite, from the late Soviet period to the Putin/Medvedev period. In the end, Chernomyrdin decided not to run for president, but he was viewed as someone with strong presidential ambitions. In March 1998, just months before the August 1998 financial crisis, Chernomyrdin was unexpectedly replaced as prime minister and removed from government and the executive branch altogether. By April 1998, he was already declaring his intention to challenge for the presidency in 2000 (Sadkovskaya 1998).

The removal of Chernomyrdin from the post of prime minister was perhaps unsurprising considering the nature of the Chernomyrdin/Yeltsin relationship.

When the idea of creating Our Home is Russia first emerged in 1995, Yeltsin completely rejected it because he believed it would significantly enhance Chernomyrdin's power. The tandem tactic of Our Home is Russia and the Ivan Rybkin Bloc was designed to allay the fears of Yeltsin and, in part, to check the ambitions of Chernomyrdin – the Ivan Rybkin Bloc meant to counterbalance Chernomyrdin and Our Home is Russia.

The dismissal of Chernomyrdin from government in 1998, as with the case of Egor Gaidar and Russia's Democratic Choice in 1994, effectively signalled the end of Our Home is Russia as the party of power. One thing all parties of power have in common is their reliance on resources dispensed from above, as preferred agents of the executive branch power-holders. Former presidential advisor Georgii Satarov (2007) emphasises this point well with the observation that 'Our Home is Russia was Chernomyrdin's clientele, but it shared the fate of every clientele – after the boss disappears, the clientele disappears'.

Without the resources and support of Chernomyrdin as prime minister, and with its largely negative associations with the Yeltsin regime, Our Home is Russia was no longer a viable proposition, either as an independent force or as the platform for Yeltsin's chosen successor, Vladimir Putin. In the end, Our Home is Russia did compete in the 1999 State Duma election but failed to pass the 5 per cent threshold and later merged with Unity in March 2000. In the face of the pressing challenge of FAR, headed by presidential hopeful and former prime minister, Evgenii Primakov, supported by Moscow mayor, Yurii Luzhkov, and St. Petersburg Governor, Vladimir Yakovlev, among others, the decision was made to create a party of power from scratch. The result was Unity, formed in September 1999, just three months before the scheduled parliamentary election. Table 2.3 shows the results of the 1999 State Duma election.

Unity was from the very outset the epitome of a party project. By late 1999, the experience of previous party of power projects, of ten years of competitive elections in the post-Soviet space and the assimilated knowledge from elections elsewhere, meant that there was a great deal of know-how to draw upon. As in

Table 2.3 State Duma election results, 19 December 1999

Party	Party-list vote (%)	SMD vote (%)	Seats (party-list and SMD)
CPRF	24.3	13.4	113
Unity	23.3	2.1	73
FAR	13.3	8.6	68
URF	8.5	3.0	29
LDPR	6.0	1.5	17
Yabloko	5.9	5.0	20
Our Home is Russia	1.2	2.6	7
Independents	–	41.7	114

Source: http://www.russiavotes.org.

1995 with Our Home is Russia and the Ivan Rybkin Bloc, Boris Berezovskii played an important role in the creation of Unity, overseeing its initial financing. However, the relationship between Unity and Berezovskii was by no means open. Berezovskii himself admitted that he had little input in the technical aspects of Unity's formation, but that he had some ideological input into the party. He also expressed his displeasure at the fact that Unity leaders, Sergei Shoigu and Aleksandr Karelin, claimed that they had never heard of him (Andrusenko 1999). Aleksandr Gurov, a third Unity leader, also claimed never to have met Berezovskii, although he did acknowledge the former's role in the formation of Unity. According to Gurov, all of his dealings with the Kremlin were conducted with presidential aid, Igor Shabdurasulov (Gurov 2007).

By 1999, and according to Andrei Loginov (2007), creating parties of power had become quite simple, with the complexity of the process reduced to three stages. In May 1994, Loginov began working for the presidential administration and later the Administrative Board on Internal Politics, responsible for parties and other movements in a role similar to that played by Vladislav Surkov in the post-Yeltsin period (see Chapter 5). In his opinion, to make a party, those in power (1) recruited good organisers and gave them their brief; (2) gave them good professionals from the security services, in order to screen potential party candidates for any unsavoury history, in particular connections with criminal groups; and (3) recruited political-technologists to run the election campaign.

Part of the Unity success was attributable to the ability of party-builders to tailor the party to the electorate, bringing resources and experience together to create a saleable brand. As discussed in the next chapter, this kind of image-conscious, media-intensive strategy is central to United Russia's electoral success in the post-Yeltsin period. The troika of leaders recruited to head the Unity party-list showed that party-builders had learnt important lessons from previous parties of power. In contrast to the relatively uncharismatic party leaders such as Egor Gaidar and Viktor Chernomyrdin, Unity deployed the energetic Minister for Emergency Situations (EMERCOM) Sergei Shoigu, Olympic wrestler Aleksandr Karelin and police colonel Aleksandr Gurov. These leaders were an important element of the unprecedented media campaign employed on behalf of Unity. The aim was to boost the party's electoral prospects and those of presidential hopeful Vladimir Putin by directly raising their profile and by discrediting the challenge of Luzhkov/Yakovlev/Primakov and FAR with negative campaigning. Political technologist Vyacheslav Nikonov, who worked on the FAR campaign in 1999, summarises the role of the media and the intensity of the struggle between the two movements:

> Unity was the weaker project, but it had government, Kremlin and TV behind it . . . I would say that these elections saw the most violent use of administrative resources, media, attacks and so on. I would consider 1999 to be the worst of all the elections we had.
>
> (Nikonov 2007)

In fact, Unity was a very deliberate and calculated project from the outset. The movement's leaders were chosen precisely to fit a predetermined campaign strategy. One of Unity's leaders, Aleksandr Gurov, recalled that, prior to the formation of the movement, he was invited to the Kremlin to discuss the possibility of his participation in the project. According to Gurov (2007), he was informed that preliminary public opinion surveys ranked him high on a list of recognisable personalities that ordinary Russians identified with, but, more importantly, he was first on a list of personalities that voters considered reliable. This persuaded key figures behind Unity of the usefulness of Gurov, who was subsequently recruited to the cause.

The figures of Sergei Shoigu and Aleksandr Karelin were also good media opportunities for Unity. Shoigu, as a government minister, often appeared on television in his official capacity, in effect providing Unity with extra airtime. The same was true of Karelin, whose inclusion in the Unity leadership was a fairly typical example of PR technology and the use of celebrity that United Russia was to employ to such good effect from 2002. All three leaders contributed to the symbolism of the Unity campaign. Unity was known as *Medved'* (bear), an image that the party used to good effect in its election campaign, while the three leaders played on the idea of three fighters (the natural calamities fighter, crime fighter and Greco-Roman wrestler), as well as the *tri bogatyrya* (three heroes) found in Russian folklore. Aside from being vigorous and energetic, all three were born outside of Moscow, which added to the overall appeal of the party. Gurov was a native of the Tambov region (Volga uplands); Shoigu was born in what is now the Republic of Tuva (East Siberia), while Karelin was born in the Novosibirsk region of Western Siberia (Mukhin 2006). This helped to create the image of an outsider party, useful for the subsequent media campaign, which aimed sharp criticism at the Moscow-based politics of the old order, including Our Home is Russia. These figures were important for other reasons too. They all had relatively clean histories with regard to *kompromat* (compromising material), which became a key part of the media attacks on Moscow mayor, Luzhkov, as the election campaign progressed. As Aleksandr Gurov recalled in his meeting with Igor Shabdurasulov in the Kremlin prior to the creation of Unity:

> So I asked, 'why us? – Shoigu, Gurov and Karelin? Karelin isn't a politician, I'm a colonel in the tax police, Shoigu isn't a politician either, why us?' He replied, 'it's very important that the first three people at the top of the party list are clean, that there isn't any *kompromat* on them'.
>
> (Gurov 2007)

The party message was also carefully tailored to deflect attention from the absence of concrete policies. In terms of electoral strategy, the appearance of the 'centrist' Unity signalled the evolution of the party of power into one that is best described as a catch-all party (Colton and McFaul 2000), a subject discussed in relation to United Russia in the next chapter. Unity in 1999 was able to avoid the negative connotations of the old Yeltsin regime, of previous parties of power and the reform process as a whole by studiously avoiding any concrete policy suggestions and by keeping the party programme to a bare minimum. Unity's programme was a little over 2,500 words long and played on certain key themes that were very much in line with voter opinion. Aside from stressing the roots of Unity in the Russian regions, the party supported the option of force to keep Chechnya as a subject of the Russian Federation, stood against corruption and bureaucracy and supported 'an effective and strong president' – all popular and relevant themes at the time (*Edinstvo* 1999).

The reversal of the principal–agent relationship

The strength of power-holders in the Russian Federation is not the main focus of this research, but, in line with the argument developed in this chapter, it is worth considering some of the sources of incumbent power that influence party of power development as part of the overall endogenous elite variable already discussed. This, in turn, gives a better understanding of the qualitative transformation of the party of power in the post-Yeltsin period and gives support to the overall argument that United Russia reverses the arrows of causality typically found between party and regime outcomes, with the strength of the party a reflection of the power of individuals above and beyond the party. These sources of incumbent power are then detailed further in the chapters that follow.

Of all the sources of incumbent power that affect party of power development, the following four appear to be the most significant. First, there is control over financial flows or the ability of power-holders to sufficiently fund party of power projects. For much of the post-Yeltsin period, the revenue from oil and gas continues to provide significant financial flows to bolster various political projects, including United Russia – building a party across a continent-sized country requires serious money. Second, there is the structure of governance in the Russian Federation, discussed at several junctures in the chapters that follow. This structure of governance is, in many respects, a legacy of the Soviet period, providing incumbents with formal and informal channels to affect political outcomes, including the electoral success of United Russia. Shevtsova identifies state bureaucracy as the most important systemic element of the regime, without which it could not survive (Shevtsova 2001: 85), and this may be taken to include the security services too. The third is state control over media outlets. In the Yeltsin period, the mass media developed into a political weapon as opposed to a traditional tool of free speech (Tret'yakov 2003: 7), while the start of the Second Chechen War in 1999 served as a useful pretext for further increasing control (Bacon *et al.* 2006). The fourth source of incumbent capacity can be seen as either an outcome of the others or a standalone resource – the personality, strength and popularity of the president in this super-presidential system.

If we look again at the success of Unity in 1999, it is possible to see each of these resources at work, although it was not until the post-Yeltsin period that they became so strong as to allow the appearance and successful development of United Russia. As mentioned previously, Unity had some serious financial backers, including Boris Berezovskii, who channelled his own money into the party project. In many ways, the cost of the Unity project was limited to a strong media campaign as there were few alternatives with so little time remaining before the elections. The use of media to boost Unity was palpable; especially the now infamous weekly news programme on the Berezovskii-owned ORT channel hosted by Sergei Dorenko, which helped derail the momentum of FAR and the Primakov/Luzhkov/Yakovlev challenge. Dorenko, nicknamed 'Berezovskii's bulldog', among other less flattering names (Zolotov 1999), aimed regular personal attacks at Moscow mayor, Yurii Luzhkov. In addition to securing media coverage, it is also highly likely that Unity's financial resources allowed for vote-buying and other forms of fraud before, during and after the election.[15]

The structure of governance was also prominent in the success of Unity. In the words of Aleksandr Gurov (2007), 'state bureaucracy played a big role for the party' and governor support was equally influential in the party's overall election results. Rose and Munro (2002: 136–7), for example, found that governor support for Unity candidates in the SMD races increased overall electoral support by an average of 9.6 per cent and support for FAR candidates increased support by 11.8 per cent. At the same time, on the party-list portion of the vote, governor support increased the returns for Unity by 6.6 per cent and for FAR by 17.9 per cent .

However, Unity itself was only a part of the larger political development of Vladimir Putin and the 'Putin factor'. Although the Unity project was simply a component of the more important goal of securing Vladimir Putin's election as president of the Russian Federation in March 2000, Unity in 1999 reflected the growing popularity of Putin. Brudny summarises this point well with his observation that Unity's success in 1999 was a 'direct consequence of the meteoric rise of Putin's own popularity and the prevailing feeling of optimism' (Brudny 2001: 168).

In the 1999/2000 election cycle, external variables, in particular changes in voter opinion, did figure in the success of Unity, but this was mainly a change in attitude towards the federal executive branch, reflected in the growing popularity of Vladimir Putin. To put this in perspective, at the start of the 2000 presidential election year Putin had a 41 per cent approval rating. At the start of 1996, Yeltsin's approval rating stood at 11 per cent (Colton 2005: 110). As such, it is the change in popular attitudes toward the president, not parties, that has been a key driver in a party of power development and its post-Yeltsin electoral success. As the next chapter demonstrates, attitudes towards parties have remained fairly constant in the Russian Federation, and even by the end of 2010 they remain one of the least trusted institutions.

Even though the Moscow apartment bombings and subsequent war in Chechnya were significant influences on public opinion in the run-up to the 1999 State Duma election, they primarily increased Putin's popularity, which in turn increased Unity's popularity. A survey conducted by the Public Opinion Foundation in March 2000 (FOM 2000) found that 45 per cent of respondents thought that the situation in Chechnya made Putin more popular, with only 12 per cent replying that it made him less popular. But war in Chechnya translated into public support only because of the ability of incumbents to manage it effectively.

The resource of Russian nationalism was something that the Yeltsin administration tried and failed to utilise during the First Chechen War, 1994–6. By the time of the Second Chechen War, the state was much better placed militarily to succeed and so power-holders were better placed to exploit public opinion.

Although assigned to the category of external stimuli, in view of the circumstances surrounding the Second Chechen War, it is difficult to ascribe it a purely contingent nature. Indeed, a case could be made that Chechnya was simply another example of the elite variable at work and another attempt to galvanise public opinion in favour of power-holders. It is worth noting that there was some considerable suspicion surrounding the Moscow apartment bombings that occurred on 4 September, 9 September, 13 September and 16 September 1999, killing hundreds of civilians and subsequently attributed to Chechen terrorists. The nature of these bombings and the so-called 'Ryazan training exercise', in which members of the Federal Security Service (FSB) were apprehended planting a substance in the basement of an apartment in Moscow on 22 September 1999 (see Satter 2003: 63–70 for details), suggests that these events and the Second Chechen War may have been orchestrated to coincide with the 1999/2000 election cycle.

Concluding remarks

Although there is continuity between parties of power, indicated by the opening quote of this chapter, the critical juncture of the 1995/6 election cycle marks a distinct stage in their development, with implications for United Russia in the post-Yeltsin period. As the following chapters demonstrate, the ability of the party to affect dominant-power politics, in line with the framework presented in the previous chapter, is hindered by its relationship with power. Rather than a principal power in the Russian political system, the party of power developed as an agent of power. This does not prevent the party from playing any role for the regime, as the following chapters demonstrate, but it does limit its value as an explanatory variable in larger regime outcomes. In effect, United Russia is misleading – its electoral success, size and dominance over the party-system simply reflect the strength of power-holders in and around the federal executive branch. United Russia is a preferred party agent of power-holders and so a reflection of their intentions and ability to project power. This theme is developed further in the next chapter, which looks at United Russia's role in managing elections in the second decade of post-Soviet politics.

3 Managing elections

Delivering the right result

Stalin said: 'The main point is not how people vote, but how the votes are counted'. So Iosif Vissarionovich was never afraid of elections, although under his regime people were afraid to vote against him, afraid that they were being watched when they voted. Today it is much easier to watch . . . they know immediately and for sure who voted for whom and for what party.

(Vladimir Lysenko 2007)

As a matter of fact, all we have are election campaigns, but not elections. Elections do not exist in Russia now. They are decorative, not real.

(Viktor Alksnis 2007)

In line with the framework presented in Chapter 1, this chapter examines United Russia's role in managing elections and delivering votes to the regime at election time – an important element in generating dominant-power politics in political systems in which elections are central. In line with the focus and substantive argument presented in this study, this chapter shows the role of the party of power in securing election success, but also the way that United Russia very much reflects the elite variable, or the intentions and ability of incumbents to make the party succeed. United Russia's election results are in no small way influenced by the ability of power-holders at federal, regional and municipal levels to utilise control over financial flows, the structure of governance, media control and the popularity of the president to boost party agents of the regime. These four factors are analysed as competing explanations for United Russia's electoral success alongside the party's own electoral potential.

The first part of this chapter reviews this electoral potential of United Russia's organisation and the ability of the party to successfully coordinate election campaigns across the Russian Federation. The first part of this chapter also considers this success in terms of a party agent enjoying a resource advantage over its competitors. The discussion then proceeds to United Russia's candidate recruitment strategy as an explanation for the party's electoral success, but also the competing explanation of the structure of governance; that the party benefits from the incorporation of candidates who have access to power and influence, independent of the party. The third part considers the effectiveness of United Russia's party message as a factor in its popularity and electoral success, alongside the decisive influence of favourable media coverage, in particular party of power bias on state-controlled television channels. This chapter concludes by considering the powerful 'Putin factor', but also the 'Medvedev factor', and how presidential popularity boosts the electoral fortunes of the party of power.

Explaining the electoral success of the post-Yeltsin regime

Overall, the ability of United Russia to manage elections in the post-Yeltsin period extends the discussion of the party of power from the previous chapter. The process of elite learning, evident throughout the 1990s, gave would-be party-builders the know-how to create an effective party organisation, one able to provide a good return on their investment and deliver on their electoral expectations. The first part of this chapter deals with the question of presidential elections and then moves on to consider parliamentary elections, in order to gauge the overall party input into the post-Yeltsin regime's electoral success.

The party of power and presidential elections

Perhaps the largest, single input made by any political party into the outcome of a presidential election in post-Soviet Russia was that of Unity in the 1999/2000 election cycle. As noted in the previous chapter, in late 1999, power-holders in and around the federal executive branch embarked on their first 'operation successor' campaign of the post-Soviet period to secure the election of Vladimir Putin to the post of president of the Russian Federation. This campaign had a backdrop of intense intra-elite conflict as regional power-holders, including Moscow mayor Yurii Luzhkov, St. Petersburg mayor Vladimir Yakovlev and their presidential candidate, former Prime Minster Evgenii Primakov, challenged the old presidential clique of Yeltsin and his 'family', including Boris Berzovskii and Yeltsin's daughter, Tatyana Yumasheva (formerly Tatyana Diyachenko), among others.

Unity was able to increase Putin's presidential election prospects by taking momentum away from FAR, party agent of Lyzhkov/Yakolev/Primakov during the 1999 parliamentary election campaign. A good indication of this momentum-halting power of Unity can be seen in the ratings of the party, August–December 1999, and the relative decline of FAR. In September 1999, at the moment when Unity was formed, FAR's rating stood at around 22 per cent – a strong position for a party sometime before the start of the 28-day campaign period prior to the election. From September 1999 onwards, Unity not only built support among voters, but also successfully reduced FAR's rating, with the party of power in waiting able to gain only 13 per cent of the vote and with Unity polling an impressive 23 per cent (Brudny 2001: 161). The more important outcome was that, nine days before the official registration for the March 2000 presidential election, Putin's strongest presidential competitor, Evgenii Primakov, officially withdrew his interest (Volkova 2000).

Despite the temptation to ascribe Unity a decisive hand in Putin's election to the post of president in March 2000, the outcomes of both the 1999 State Duma election and the 2000 presidential election reflected the significant elite manoeuvring that was taking place behind the scenes. As noted in the last chapter, the rapid rise in Putin's popularity from August 1999 boosted the popularity of Unity during the parliamentary election campaign, but a no less significant event was Yeltsin's unexpected resignation on New Year's Eve 1999, and the subsequent elevation of prime minister Putin to the post of acting president. This promotion provided Putin with a decisive platform from which to secure his presidential ambitions less than three months later. It should also be noted that the 1999/2000 election cycle was characterised by the genuine strength of the opposition challenge, and no presidential election since has entailed so much uncertainty. As a result, and despite the merger of Unity and FAR in 2001 (detailed in Chapter 5), United Russia's influence over presidential elections remains minimal, although Chapter 4 considers the party's indirect input in the 2008 presidential election when it played the role of institutional makeweight to allow Putin to step down as president. Although Vladimir Putin (2004) and Dmitrii Medvedev (2008) were officially nominated by United Russia's Congress to represent the party and run for the post of president, on both occasions this was simply a rubber stamp of a decision made elsewhere, the party leadership and congress entertaining no real debate or control regarding this matter.

The problem for Russian parties in terms of influencing presidential elections is that electoral law affords them no monopoly over candidate selection and nomination for this most powerful executive position. According to federal law, parties have the right to nominate presidential candidates, but candidates may also self-nominate and thus run for presidential office without the support of any party, further weakening their relevance in this respect. In effect, this is a legal provision allowing independent candidates to run for the highest executive office and, to date, no president has been party affiliated beyond an informal association. President Yeltsin (1991–9) and Putin (2000–8) were not party members, and the same is true for President Medvedev (2008–12).

Overall, the value of the party rubber stamp for presidential candidates reveals the instrumental uses of parties, in particular parties of power, in post-Soviet Russia. Their practical value is that they may help a candidate to collect the substantial signatures required to run in the presidential election, which according to the law 'On Elections of the President' stands at no fewer than two million, with no more than 50,000 signatures allowed from any one subject of the Russian Federation (CEC 2010a). In the 2008 presidential election, only two of the six candidates who attempted to run for this office were self-nominated (Andrei Bogdanov and Mikhail Kas'yanov), with Kas'yanov eventually denied registration on the grounds that too many of the 400,000 signatures checked by the CEC were either falsified or unable to be authenticated (CEC 2008a).

In addition to this practical consideration of signature collection, party nomination of presidential candidates presents a watered-down version of a party president to voters who need convincing that parties actually serve a constructive purpose in Russia. As discussed later in this chapter, this also helps establish the 'coat-tail effect', important for subsequent party election success, by associating the party with a popular presidential candidate. Party support for a non-party presidential candidate may also satisfy rank and file supporters that they are in some way involved in power politics at the highest level, and so keep the party faithful motivated – a by no means redundant consideration for the party of power.

As a final point, and just to underline their insignificant input into managing the outcome of presidential elections in Russia, changes to the Constitution in 2008 are likely to further reduce the role of parties. The timing of the December 1999 parliamentary election just before the more important presidential election in March 2000 created an extended campaign period, in which the former acted as a primary for the latter (McFaul 2001: 1167) and, to a lesser extent, the same kind of dynamic can be seen in every federal election cycle, 1995/6–2011/12. On 21 November 2008, the State Duma passed amendments extending the length of presidential office from four to six years and for legislators from four to five years (Bilevskaya and Rodin 2008). This means that national presidential and parliamentary elections will fall on different years after the 2011/12 cycle, lessening the impact of parliamentary elections on the outcome of presidential elections by removing the primary element.

Parliamentary elections in the post-Yeltsin period

It would appear that, even if United Russia has limited input into presidential elections, its success in parliamentary elections has contributed significantly to dominant-power politics – a circumstance when one political grouping dominates, despite the presence of democratic institutions, including elections. The results of the 2003 and 2007 State Duma elections, for example, show that United Russia mobilised pro-government and pro-president support with a level of success unparalleled in the post-Soviet period. To put this in perspective, in the 1999 State Duma election, Unity and FAR accounted for 24 per cent of the overall vote, translating into 41 per cent of Duma seats. In the 2003 State Duma election, the newly formed United Russia surpassed the combined Unity/FAR total, gaining 30 per cent of the vote and 49 per cent of Duma seats. In the 2007 State Duma election, United Russia achieved something akin to a super-majority, collecting 64 per cent of the vote and 70 per cent of seats in the national legislature. Table 3.1 shows the top three party performers at each of the five State Duma elections (1993–2007), confirming United Russia as the most successful party in post-Soviet Russia to date.

This electoral success is part of the overall puzzle of United Russia identified in the introduction chapter, which seemingly contradicts a great deal of literature on party development in the Yeltsin period (Bacon 1998; Enyedi 2006; Evans and Whitefield 1993; Fleron and Ahl 1998; Ishiyama and Kennedy 2001; Kopecky 1995; Lewis 2000; Toole 2003). The question of interest is whether this electoral

Table 3.1 Top three party performers in State Duma elections, 1993–2007

Party winners (list/SMD) (%)				
1993	1995	1999	2003	2007
LDPR (21/3)	CPRF (22/13)	CPRF (24/13)	United Russia (38/23)	United Russia (64/–*)
Russia's Choice (14/6)	LDPR (11/5)	Unity (23/2)	CPRF (13/11)	CPRF (12/–*)
CPRF (12/3)	Our Home is Russia (10/5)	FAR (13/9)	LDPR (11/3)	LDPR (8/–*)

Source: http://www.russiavotes.org.

*The SMD portion of the vote was abolished for the 2007 State Duma election.

success in the post-Yeltsin period is really a reflection of United Russia's efforts or of other, non-party factors.

Although electoral success is a complex combination of party efforts (Farrell 2002: 64), this chapter discusses three aspects in particular, in order to ascertain United Russia's actual role in managing elections. These aspects include the party's organisational input into election success, the party's candidate selection strategy and the party's overall campaign message to voters. As for the first aspect – party organisation – it is immediately obvious that United Russia, in comparison with previous parties of power, is able to utilise an extensive physical network of branches and members, which represents an immediate resource for the party at election time. As already noted, figures credit United Russia with 2,597 regional branches and 53,740 local branches, together with a membership base that, according to the Ministry for Justice, numbers 2,073,772 by February 2011 (Ministerstvo Ustitsii 2011a).

At a deeper level, the advent of organisational complexity necessitated the professionalisation of the party of power and this is another aspect of United Russia that distinguishes it from its predecessors in the Yeltsin period. United Russia departs from previous party of power experience in the quality and number of full-time party workers and in the party's managerial expertise, which, overall, contribute to the effective coordination of party activities. Effective coordination, as a component of electoral success, includes the ability of the party to ascertain the state of the electoral market, deploy party resources when and where needed across a huge territory, synergise know-how within its organisation and develop the all-important party brand for voters.

In the first instance, ascertaining the electoral market means collecting intelligence on voter attitudes as well as information on the relative strengths and weaknesses of the party vis-à-vis party competitors. United Russia, unlike previous parties of power, employs full-time workers to collate public opinion data and to supplement established public opinion forecasters, such as the Public Opinion Foundation (FOM) and, in particular, the Russian Public Opinion Research Centre

(VTsIOM), which works closely with the party. As discussed later in this chapter, data detailing trends in public opinion represent an important component of the party's overall electoral strategy. Public opinion is also crucial from the perspective of dominant-power politics, as it provides important indicators of levels of popular dissatisfaction with power-holders. So, the work of United Russia serves a dual purpose: it furthers the party's overall electoral ambitions and provides an early warning system for the regime by flagging emerging public discontent.

However, effective coordination is evident not only in the party's ability to accurately gauge public opinion, but also in its ability to act on this information. Once the party has established a clear picture of the electoral landscape, regions in need of specific or particular assistance can be highlighted. This then forms the second aspect of campaign coordination, which includes the allocation and delivery of resources essential for electoral success.

The need for United Russia to prioritise regions during election campaigns relates as much to Russia's unique electoral geography as to the finite nature of resources available to any party. As well as being expansive, covering eleven time zones, the Russian Federation is quite heterogeneous in terms of population density. According to official state statistics, over a third of Russians live in only ten regions of the Federation.[1] As such, the party's success in both the 2003 and 2007 State Duma elections was in no small part a result of its ability to identify population centres that required more intensive party campaigning, and then to allocate resources appropriately.[2] Opinion polls taken prior to the 2003 State Duma election, for example, showed that United Russia received higher levels of support from voters living in smaller settlements averaging between 50,000 and 100,000 inhabitants, while the lowest levels of party support were found in towns and cites with a population of 500,000–1 million or more (Yanbukhtin 2008: 47). Consistent election success across the regions requires significant resources and a well-managed organisation in order to effectively concentrate party efforts where they are needed most.

Perhaps the greatest strength of United Russia in comparison with previous parties of power is seen in its development as a truly all-national party, a subject discussed in more detail later in this chapter as well as in the following chapters. This development again underscores the need for capable party management able to incorporate disparate know-how across such as large organisation. In many cases, regional offices and party candidates are given licence to tailor election campaigns to meet specific local conditions, sometimes employing their own intelligence on the electoral market through use of focus groups, opinion polling, etc. As one party official commented:

> We invite party managers from the regions to participate in training events held on various subjects, such as election campaigning or organisational work. So, we constantly work with them. As a result, there is no need for them to consult with us in their everyday work.
>
> (Il'nitskii 2007)

Local knowledge is extremely important for United Russia's electoral success and so the ability of the party to secure election victories at every level depends on the successful incorporation of this knowledge into the overall party election campaign. The variation in culture and language across the Russian Federation, in particular in the national republics, has a large bearing on the effectiveness of PR technologies employed during election campaigning. The technique of direct mailing to voters, for example, entails certain risks when used in those Russian regions with what may be described as 'eastern cultures', where high-status candidates may lose the respect of voters if they correspond directly with ordinary civilians. Again, the complexity of Russian electoral geography makes campaign coordination essential for party success. A specialist at the political consulting agency, Niccolo M, summarises this point well:

> It is very important for a person who studies Russia to know for sure that a unified Russia is a myth. It does not exist. In reality, Russia is a very heterogeneous space made up of very different political and cultural-political areas.
>
> (Radkevich 2007)

Effective coordination of party activities is also important for developing the party label or brand. As previously noted, the skilful use of media figured prominently in the success of Unity in 1999 and continues to be an essential component of party of power success in the post-Yeltsin period. Aside from more traditional means of voter mobilisation, such as door-to-door campaigning, and other face-to-face methods that United Russia has employed in recent years, the party has also used its organisational presence in the regions to increase the party's media exposure. For example, party projects in the form of all-national *konkursy* (competitions), aimed at raising awareness of certain social issues, provide the party with important PR opportunities. Party initiatives include *Nash Gorod* (Our Town), *Nashi Roditeli* (Our Parents) and *Biblioteki Rossii* (Libraries of Russia) among many others, designed primarily to ensure that the party is continually in the media spotlight.

The party's VII Congress, held in December 2006, approved a number of new projects, including *Sochi 2014*, which has a mission statement to 'build new sports infrastructure' ready for the Winter Olympics in 2014. The project had a reported 327 billion rouble budget, although the exact role of the party in either securing or spending this money is unclear (Edinaya Rossiya 2009a). The important point is that, during the short but intensive election campaign periods, voters inevitably experience an information overload, so only large-scale, memorable news events clearly differing from the norm, such as these projects, are likely to make a positive and lasting impact.[3] By the end of 2010, the party had almost 60 ongoing social and infrastructure projects.

In addition to providing opportunities for publicity, these *konkursy* and similar party initiatives are important for other reasons. They allow the party to stress its real achievements in improving the quality of life for ordinary Russians, supporting the party slogan *partiya real'nykh del* (the party of real achievements), used consistently throughout the period 2002–10. This slogan supports the image of United Russia as the party of the actual moment, working for the good of the country, in comparison with the party of the past (communists) and parties of the future (those liberal leaning, pro-Western parties that promised a better life for Russians in the 1990s, without delivering). The large and visible party organisation goes some way to affirming the image of the party as an important and active feature of the political landscape, distinguishing the post-Yeltsin party of power from its predecessors. The next chapter considers the actual achievements of United Russia in more detail.

Resource advantage and electoral success

The professionalisation of the party of power was accompanied by its development as a truly all-national party in the form of United Russia, but all parties need resources to build and maintain their organisations, run election campaigns, attract supporters and perform the numerous functions they are typically ascribed. In this respect, the organisational strength of United Russia very much relies on the party's continuing status as the preferred agent of the federal executive branch. As demonstrated in the material that follows, this fact alone questions the ability of the party to attract substantial electoral support independent of its more powerful sponsor. Overall, the resource advantage enjoyed by the party is staggering, showing how the federal executive branches' control over financial flows, especially the revenues from the sale of oil and gas, feeds into political success and stability. These resources strengthened the bargaining position of the federal executive branch vis-à-vis the regions, encouraging regional administrations to adopt United Russia as their preferred party agent too – a key part of the centralisation process (see Chapters 4 and 5).

United Russia is different to its predecessors in this respect, enjoying superior financing, although parties of power have (unsurprisingly) outperformed their party competitors fairly consistently in this respect throughout the post-Soviet period. In 1993, it was widely understood that Russia's Choice enjoyed a financial advantage over rivals, although party leader, Egor Gaidar, claimed that his party's resources were at best on a par with those of Yabloko (Gaidar 2007). However, the official figures published by the CEC show that the party spent over 1,500,000 US dollars on its 1993 election campaign, some way ahead of the second-placed party, PRES, which spent 669,000 US dollars (Treisman 1998b: 4).

In a similar way, in 1995 Our Home is Russia was widely acknowledged to have significant financial backing, although the party tried very hard to appear unexceptional in this regard, with the kind of creative bookkeeping that makes this subject extremely difficult to ascertain with any certainty. In November 1995, the party returned a financial statement to the CEC showing that it had collected 80,000,000 roubles (1,700,000 US dollars) in campaign accounts, but had not spent a rouble during the campaign (Stanley 1995). To put this in perspective,

LDPR's campaign accounts for the same period showed that it had collected 9.9 billion roubles (21,000,000 US dollars) and spent 4.5 billion roubles (9,000,000 US dollars). Belin and Orttung (1997: 69–70) note that, even though Our Home is Russia was endorsed by many banks and had an extensive billboard and television campaign, the gas giant Gazprom, which party leader Chernomyrdin headed from 1989–92, was not officially listed as a campaign contributor, raising questions as to the real extent of party campaign financing. As for Unity, in 1999 there were rumours of campaign resources totalling more than 30,000,000 US dollars (Latynina 2000), although without insider information it is impossible to substantiate many of these claims. Party financing remains a subject that few party elites in Russia are prepared to discuss.

As for United Russia, official figures from 2002 (the first year of the party's existence) showed that the party had successfully accumulated more donations from organisations and companies than any other party, attracting almost 400,000,000 roubles, around four times more than the second-placed People's Party with 115,000,000 roubles (Tsygankov 2003). These figures also showed United Russia to be extremely economical, having the lowest expenditure of any party for that tax year, apparently spending nothing on congresses, conferences, television commercials, publications or printing. By the end of the decade, the situation had not changed greatly and, if anything, figures show United Russia to have consolidated its financial dominance over its party competitors. According to the CEC, United Russia received just over 270,000,000 roubles in the last quarter of 2010 (just over 9,000,000 US dollars), almost twice as much as the combined income of (151,700,000 roubles) of the remaining six registered parties (CEC 2010b).

In the case of United Russia, sources of funding reveal something of its hybrid nature and the real mobilisation potential of the party. United Russia, as the previous discussion indicated, has increasingly acquired the characteristics of a mass-party, including organisational complexity, a large party membership and, by suggestion, a large pool of activists ready to support the party. However, the impressive size of United Russia's membership must be placed in the context of the dimensions of the Russian Federation, as well as the nature of party funding. Although impressive by post-Soviet Russian standards, its two million members at the time of the 2007 State Duma election only equated to 1.8 per cent of the voting age electorate (for details on the size of the Russian electorate see IIDEA n.d.). This figure is still lower than party membership in many Western democracies, which has itself experienced consistent, long-term decline since the 1960s.[4]

At the same time, the party actually relies little on member subscriptions as a source of support. This is underlined by the decision taken at United Russia's X Congress in November 2008 to change the basis of member subscriptions from mandatory to voluntary (Edinaya Rossiya 2009b). However, even before this decision, United Russia's member contributions were largely symbolic in character, accounting for no more than a twentieth of the party's official expenditures in the period 2004–8, compared with just under a sixth for the CPRF.[5] Table 3.2 provides a breakdown of party revenue and expenditures, including the income generated by member subscriptions, in the period 2004–7.

Table 3.2 Party finances, 2004–7 (millions of roubles)

	Donations	Expenditure	Membership fees
United Russia	3,200.4	4,100.8	209.9
LDPR	701.7	726.3	2.6
CPRF	434.5	609.7	110.6
URF	424.1	457.6	0.02
Yabloko	238.2	264.5	0

Source: CEC (2009a).

Overall, party financing is important for understanding United Russia's status as a genuine mass-party. In historical terms, the difference between mass-parties and other party types relates to their respective sources of funding. According to Duverger, recruiting members is a fundamental activity for the mass-party, as subscriptions paid by party members form a crucial financial resource. The cadre party, in comparison, may imitate the mass-party by instigating a mass membership, but its essential characteristics are that of a group of notabilities or influential persons whose name, prestige, connections and electoral expertise help secure votes (Duverger 1964: 63–4). In post-Soviet Russia, it is only the CPRF that raises a significant proportion of its budget from member fees. From this perspective, United Russia is a mass-party only in the sense that it needs to integrate across a large territory, not in the sense that it has an active membership.

So, in the vein of cadre parties elsewhere, the majority of Russian parties, including United Russia, rely on donations from wealthy individuals as well as the enterprises they control to complement the funding that every party receives from the state. A survey of United Russia's accounts reveals the scale of the party's 'official' funding, as well as the web of interconnections between economic interests and power. In the 2009 tax year, 57 companies of various descriptions donated sums from 445,000 to 32,642,000 roubles (approximately 15,000 to 1,120,000 US dollars). One interesting feature of United Russia's accounts is the lack of serious donations from big business, including Russian oil and gas companies, which reveals something of the lobbying potential of the party (discussed in Chapter 4) as well as the kind of selective benefits the party is able to offer candidates (discussed in Chapter 5). In short, the party offers little for larger companies in terms of political influence, whereas smaller and medium businesses may offer financial support to the party as a form of insurance, in which there is a perception that the party may help overcome administrative barriers and the threat posed by organised crime.

The building company, Kavkaztsement, is a fairly typical example of one of the larger companies 'investing' in United Russia, donating 30,000,000 roubles (just over 1,000,000 US dollars) in 2009. For most of the post-Yeltsin period, this company has maintained strong connections with both United Russia and the Karachaevo-Cherkessia Republic administration, showing how the party benefits from the support of key actors in the regions. In 2004, an article appeared in *Profil'* magazine suggesting that the president of the Karachaevo-Cherkessia Republic

(2003–8), Mustafa Batdyev, 'practically owned' *Kavkaztsement* (Braginskaya 2004), while as of 2010 the chair of United Russia's regional Control-Revision committee for the Karachaevo-Cherkessia Republic, Ol'ga Prygunova, is also the deputy director of *Kavkaztsement* (Edinaya Rossiya 2011a).

Aside from business donations, the party also receives donations from individuals – 103 making formal donations in 2009, with sums ranging from 21,000 to 1,000,000 roubles. Federation Council Senator, Amir Gallyamov, for example, donated 1,000,000 roubles, just over 30,000 US dollars, which according to the senator's official financial declaration for that tax year (Shtykina and Barakhova 2010) was almost a third of his 2,700,000 rouble (93,000 US dollars) income. Regional funds represent the final official source of party income, created to engage in fund-raising on behalf of the party across the territory of the Russian Federation. The party's official accounts submitted to the CEC for 2009 showed that 71 regional funds donated amounts ranging from just over 60,000,000 roubles (Vladimir Regional Fund) to just over 1,000,000 roubles (Murmansk Regional Fund) or between 30,000 and 2,000,000 US dollars. Again, these are official figures and the actual extent of funding is likely to be much greater.

Financial resources are extremely important for United Russia's overall electoral success. As discussed in the previous chapter, from the 1999 State Duma election, the party of power developed a catch-all election strategy employing a fuzzy programmatic focus to reduce the risk of alienating segments of the electorate with concrete policy promises. Instead of attempting to persuade or sell the electorate a particular ideology, as was arguably the case with Russia's Choice and Our Home is Russia, the message was carefully but superficially tailored to resonate with as many voters as possible. In 1993, Russia's Choice crafted its election programme around the core principles of freedom, property and rule of law, while, in 1995, Our Home is Russia offered a more practical programme, but still one that persevered with the liberal democratic impulse of the early 1990s (Korgunyuk and Zaslavskii 1996). In contrast, Unity and United Russia have succeeded with single ideas or sound bites, such as support for a 'unified Russia', for 'stability' and for 'Putin' and the 'established course'. For such a catch-all strategy to be successful, the party of power needs not only access to reliable intelligence on the electoral market to tailor its message, but also sufficient media exposure between elections in order to develop the party label or brand. Financial resources enable the party to increase its media exposure, to employ professional consultants and train party workers to use the media effectively. As such, United Russia's electoral strategy may also be described as 'remote' (Toole 2003: 110) in the sense that the party is reliant on media usage to convey the party message.

Candidate recruitment and voter mobilisation

United Russia's organisational characteristics show the party to be a hybrid or mixture of styles, including catch-all and cadre party, but in line with the preceding discussion in Chapters 1 and 2 it is the origins of the party as a controlled party agent that have an enduring influence on the party's ability to effectively manage elections. Over time, the party of power developed from an elite party with some roots in society (seen in Russia's Choice at the time of the 1993 State Duma election) to a party that was much more top-down in nature and strictly subordinate to the federal executive branch (as with Our Home is Russia and Unity). As a preferred party agent of the federal executive branch, United Russia enjoys considerable advantages over its competitors, which affect the kind of electoral success the party enjoys.

These advantages are not confined to superior financing, but touch on the overall structure of governance in the Russian Federation and the nexus between state and society that translates into election success. In the city of Krasnodar, for example, the United Russia office is housed in the building of the Ministry of Transport and this pattern is no doubt repeated throughout the Federation. In addition, many party workers and party members are employed by the state in some capacity, and many state-run organisations continue to work closely with United Russia. In essence, federal law forbids civil servants from participating in politics (see Chapter 4) although the reality is often far from this ideal, as the case study of the municipal election detailed in this chapter demonstrates. United Russia has contacts with a number of state-run organisations including VTsIOM, as already mentioned, but also the Russian Academy of Civil Servants. The origins of the party of power, discussed in the previous chapter, make the relationship between state bureaucracy and United Russia very close.

Candidate recruitment and electoral success in the regions

This point, of the party benefiting from administrative resources, is amplified when viewed from the perspective of the party's candidate selection strategy. Although United Russia offers potential candidates and members a range of selective and collective incentives, a theme discussed in more detail in Chapter 5, there is another side to this relationship. In line with the previous chapter, which identified the structure of governance as a factor in the increasing strength of incumbents in the 1990s, candidates often bring their own extra-party resources to United Russia's election campaign efforts. This section examines the extent to which United Russia has succeeded in incorporating institutionally affiliated elites into its organisation and how their inclusion enhances the party's overall electoral prospects.

Power-holders at every level of governance in the Russian Federation have access to resources that are of great benefit to political parties. The research of Golosov (2004), Hale (2006) and Smyth (2006) convincingly demonstrates how, before the appearance of United Russia in late 2001, incumbents holding positions in regional and municipal power structures utilised their institutional affiliation to boost their own electoral prospects, without the need for party labels or other party resources. In Russia, the power of institutional affiliation is enhanced by specific post-Soviet conditions linked to the legacy of the Soviet command economy and the persistence of patrimonial communism, touching on issues of welfare provision, the existence of mono-towns, etc. (Hale 2003; Kitschelt 1995: 453; Lynch

2005: 128–65). As such, institutional affiliation not only includes office in the formal structures of governance, but also positions of authority in other important organisations, such as state- and privately owned enterprise, where control over resources becomes a source of power in its own right.

The relationship between United Russia and regional elites, or, more generally, regional administrations, is an interesting and important one that relates the effects of candidate recruitment to the ability of United Russia to successfully manage the outcome of elections. United Russia's strategy of incorporating well-connected regional elites, it should be noted, has been an unparalleled success from the party's inception. The merger of Unity and FAR provided the emergent United Russia with several long-standing and high-profile regional leaders, including the former mayor of Moscow, Yuri Luzhkov (1992–2010), and the former president of Tatarstan, Mintimer Shaimiev (1991–2010), both of whom were elected to co-chair United Russia's Higher Council during the party's third Founding Congress in December 2001. After the merger, the party's II Congress held in March 2003 signalled the start of a further engagement with regional leaders, with another six governors recruited to the party's Higher Council (Petrov 2003). By the time of the December 2003 State Duma election, there were clear indications that this party of power was to be a genuine synergy of regional- and federal-level incumbent capacity, with just under a third of all regional governors (28 regional leaders in 88 regions) running on United Russia's party-lists (Mereu 2003).

Aside from the inclusion of regional executives on the party's Higher Council and party-lists, the extent of United Russia's incorporation of institutionally affiliated elites can be seen in the composition of the party faction in the State Duma. An analysis of the previous occupations of United Russia deputies showed that, by the Fourth Duma Convocation, 2004–7, almost half the faction contained deputies whose previous occupation before taking up their mandate had been in either federal or regional state employment or in private business. A total of 74 of the 303 United Russia deputies came from state institutions in one form or another. These state institutions mostly comprised regional executive and legislative positions, such as governors, regional Duma deputies and members of regional government, with some additional deputies having previously worked in the Federal Executive, Federation Council and security structures, including the FSB. Another 34 United Russia deputies previously held positions in private business, mostly directors and owners of enterprises, while 38 deputies held senior positions in state-run enterprises.[6]

On the whole, candidates with business connections, both private and state-owned, represent an important resource for the party. This is true in relation to the party's formal reliance on donations, but also in relation to the previously mentioned structure of governance, where the persistence of Soviet-era economic infrastructure often gives business elites control over resources useful for clientelism, such as control over the allocation of housing, childcare and even health care (Golosov 2004: 152). These candidates with business backgrounds, in turn, have their own mobilising potential, especially for voters apathetic to parties, but who perceive some future utility in casting their vote for their own local elite.

In 2007, the journal *Finans* reported that 20 of the 500 richest Russians were State Duma deputies, 19 belonging to the United Russia faction (Smirnov 2007). Included in the ranks of United Russia US-dollar billionaires were Andrei Skoch, former entrepreneur who joined the United Russia faction in 2003 and who subsequently found his way to the Duma Committee for Industry, Construction and High-Tech Technologies. Another billionaire and well-travelled political figure, Aleksandr Lebedev, originally joined Our Home is Russia in 1997 before successfully gaining election to the State Duma in 2003 on the Motherland (Rodina) ticket. Although a vocal critic of the then Higher Council co-chair, Yurii Luzhkov, Lebedev switched to the United Russia faction during the Fourth Duma Convocation, before switching again to A Just Russia prior to the December 2007 election. He was unsuccessful in his attempt to make the Fifth Duma Convocation with A Just Russia and was later removed from the party's leadership in April 2008. In 2010, *Forbes* published their 2009 'power and money' rating for Russian bureaucrats, deputies, senators and managers of state corporations (Terent'ev 2010), calculating that 60 of the 100 richest individuals on this list were members of United Russia.

Another indicator of the predominance of institutionally affiliated candidates within the ranks of United Russia can be seen in the occupation of party secretaries at the regional level (Figure 3.1). Analysis of the background of United Russia's 84 regional secretaries just before the 2007 State Duma election showed that 72 simultaneously held positions in either the executive or legislative branches of that region or the State Duma. As of 1 September 2007, 15 secretaries held positions in regional executives, 37 in regional legislatures, 16 in the State Duma and four in the Federation Council. The remaining 12 party secretaries held prominent positions in their respective regions, including a Chief Federal Inspector, a regional

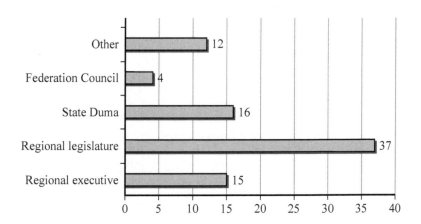

Figure 3.1 United Russia's regional secretaries by occupation, 2007. Source: Edinaya Rossiya (2007b).

Trade Union Chair, a head of Municipal Education, factory and bank directors (3) and heads of academic establishments (3).

The effects of institutional affiliation

The advantages of incorporating institutionally affiliated elites, including business elites, are numerous. As mentioned, high-profile candidates often have the kind of name recognition that may win over voters who are unsure of the party itself. This name recognition has the ability to create a coat-tail effect to the benefit of the party as a whole. Coat-tails have been well documented in American politics, where popular candidates at the top of party-lists carry other, less popular candidates lower down the list into office (Samuels 2002: 468). As for Russian politics, Brudny referred to coat-tail effects as far back as the 1993 State Duma election when Yabloko candidate Boldyrev boosted the party's results in St Petersburg on the party-list portion of the vote as a result of his own personal popularity in the city (Brudny 1998: 362).

In the Russian context, the popular candidate at the top of a party-list is termed a 'locomotive', with the ability to boost the party's mobilisation potential, and this election strategy has clearly figured in United Russia's campaigns throughout the post-Yeltsin period. By the 2007 State Duma election, 65 governors ran on the United Russia ticket, 63 heading the regional party-lists, with the remaining two occupying second place on the list (Kynev 2007). However, confirmation that this was a deliberate locomotive strategy is found in the fact that few of these governors took up their mandate in the State Duma after successful election.

Tkacheva (2008) found that the proportion of governors running on party of power party-lists who declined their mandate increased almost ninefold between 1999 and 2007. In 1999, a total of eight governors headed the party-lists of Unity, FAR and Our Home is Russia, with only one governor accepting his seat after the election. For the 2003 State Duma election, of the 28 governors that were placed at the top of United Russia's party-lists, not one accepted their mandate after the election. In 2007, of the 65 governors running on United Russia party-lists, only one actually took up their seat in parliament. This pattern is repeated for elections to regional assemblies in the same period (Tkacheva 2008: 22).

Aside from the personal resource of name recognition, the inclusion of local notables and, more generally, institutionally affiliated elites provides the party with access to other resources controlled by these candidates, often termed the administrative resource, so effective in influencing the outcome of elections. This is by no means an easy area to explore, either conceptually or empirically, but requires a deeper understanding of the electoral processes in the Russian Federation. To this end, it is useful to briefly consider some of the non-competitive or sub-competitive aspects of elections in this context, before considering a case study example.

As noted in Chapter 1, elections in Russia hold a great deal of significance for all involved, even if subject to widespread manipulation. However, it is the very possibility of manipulation that makes elections in this kind of system a defining feature. The rule of law is an essential component of a functioning democracy, as democracy survives only if officials have incentives not to intervene in the outcome of elections (Weingast 2003: 110). Gel'man (2004), developing the work of North (1990), notes that the presence, and also the absence, of the rule of law depends on the prevalence of two kinds of institutions, which are, in fact, the only two kinds of institutions that can exist. These institutions are universal (formal) institutions and particularistic (informal) institutions (Gel'man 2004: 1021). In the Russian case, the persistence of informal institutions weakens the rule of law and, in a political context, allows incumbents a degree of impunity that their democratic counterparts rarely enjoy. The rule of law in Russia is not completely absent, but assumes what Gel'man terms a 'fuzzy legality' (Gel'man 2004: 1030).

The incorporation of so many institutionally affiliated candidates provides United Russia with access to many of those particularistic, informal institutions that feed into the party's electoral performance. The following material presents a study of a district election in Moscow in 2004, to illustrate the way that incumbent strength in the regions benefits United Russia and the way that control over informal institutions contributes to the party's electoral success. This case study is based on interviews conducted in 2007 with local residents and an independent candidate who participated in the aforementioned district election in Moscow. The interviews are supported by newspaper reports from both local and national press. For ethical reasons, names of interviewees and place names are not provided.

Moscow City, under the administration of former mayor Yurii Luzhkov, represents a good example of a region where incumbent strength is high. Luzhkov was first appointed mayor by Boris Yeltsin in 1992, going on to win three consecutive mayoral elections, the last of which was in 2003. Before his removal from office by President Medvedev in September 2010, Luzhkov ruled Moscow for the best part of 20 years. Luzhkov, as a political figure with strong name recognition among voters, has also played the role of locomotive for United Russia. In December 2003, he was elected to the State Duma on the United Russia ticket, but declined to take up his mandate. The Fatherland movement that joined forces with All-Russia to challenge Unity in 1999 was essentially a regional party of power serving the interests of Luzhkov and his administration (Makarenko 2000). Luzhkov's business interests include the Sistema conglomerate, described as a 'financial and industrial empire', controlled as a branch of the Moscow government but operating in the private sector (Freeland 2000: 159).

The municipal elections held in Moscow in the spring of 2004 were in response to the law 'On Local Self-Government for Moscow', adopted in 2003, according to which each of the 125 districts of Moscow was required to form local committees comprising 12 local candidates. The campaign to ensure that United Russia candidates dominated these positions provides a good illustration of the way in which control over resources, including informal institutions, has the potential to influence electoral outcomes. As mentioned, the information on these elections comes from local and national newspaper reports, but primarily from interviews conducted with an independent candidate and local residents in the district in question. The following account is separated into three stages: before the election, during the election and after the election.

Before the election

Electoral support for United Russia in the run-up to these Moscow district elections in 2004 was enhanced by the existing strength of the local administration, as is the case in many other regions across Russia. In the case of Moscow, patronage networks used to create and sustain support for the Luzhkov administration are placed at the disposal of United Russia. For example, distributing free sausage to the local War Veteran's organisation on important holidays is a common method used to build popular support, repeated in districts across the city. The local factory supplying the sausage may be owned by an aspiring businessman who has already joined United Russia, or, if not, by someone who realises the importance of distributing gifts for the success of his future business and/or political career. In the case of Moscow under Luzhkov, free sausage does not need any personal calling card; the gift is understood to come from the local administration, which means Luzhkov and, by association, United Russia.

Just before the election and according to the account of the independent candidate, a second more active stage of intervention began. The local election commission was instructed to ensure that, of the 12 representatives required for the district, four were United Russia candidates and four were representatives of administrative bodies within the district. Formally, candidates from administrative bodies are independent of party control, typically holding positions of prominence, such as the head of local communal services or the chief doctor of a local hospital. In reality, as state workers or so-called *byudzhetniki* (those employed by the state), these people are dependent on the local authorities, not least for their job security. These local authorities form part of the hierarchy of power controlled by the Moscow City administration at the top, which in turn supports the federal executive branch and their preferred party agent, United Russia. To maintain a degree of legitimacy for these elections, the remaining four candidates were to be independent citizens of the district.

During the election

During the election campaign, efforts were made to ensure that United Russia held the upper hand over its party and independent candidate competitors. Workers at the local communal services were allegedly instructed that their main duty during the election campaign would consist of removing street advertisements for all other candidates than United Russia's, with workers threatened with pay deductions if they did not comply. Days before the election, district juniors from military schools were deployed to stand watch in front of each five to ten blocks of flats and 'control' the area so that only United Russia's leaflets and advertisements were on the walls. There were also reports that *byudzhetniki* – notably kindergarten workers – were instructed to vote for United Russia or face uncertain consequences, usually understood as dismissal. There were also reports that in the actual polling station, voters were intimidated by 'independent' observers in a final effort to influence electors.

After the election

In March 2004, in the Moscow district in question, independent candidates caused an upset and were successful in returning seven of the 12 candidates. The local election commission, in the presence of several independent candidates, counted the votes, and all observers signed a protocol confirming the final result. However, these results were later revised upwards in favour of United Russia. What is interesting in this particular example is that one of the independent candidates noticed a light on late at night in the local administrative building where the votes had earlier been counted and upon investigation discovered an attempt at falsifying the vote in progress. The building where the votes were counted was the local administrative building (*uprava*), which, as part of the administration's offices, contravened electoral law, as elections to legislative bodies cannot be organised in the buildings of executive bodies. The police were called and the story made local and national headlines. In the aftermath, a court agreed that a re-election should be held. The local police chief, who supported the claim of falsification, lost his job.

The second election campaign proved particularly difficult for the independent candidates. They were denied access to local television despite every registered candidate having an entitlement to free television time. The firm printing their campaign leaflets stopped work after receiving pressure from the authorities. The sanctions available to local administrations to punish local businesses are numerous, but a popular method is the threat or actual visit from the local fire inspector, who has the power to close buildings with immediate effect if they are found in violation of city regulations. The fire inspector, as a state worker, is susceptible to pressure from above and so forms just one link in the chain of administrative resource. In the face of these obstacles and a number of threatening phone calls made to the homes of several independent candidates, the second election saw United Russia gain the working majority in the district in question, which had been planned from the outset.

Legal provisions, in particular existing electoral law, should prevent many of the abuses documented above, but in this case informal institutions were stronger than the formal institutions designed to safeguard the competitiveness of elections. This is not to say that based on this example one can conclude that the entire electoral process in Russia is flawed or that, in the case in question, either Luzhkov or the United Russia leadership actively participated in electoral fraud (they did not need to). This example simply shows how United Russia's inclusion and incorporation of institutionally affiliated elites with control over administrative resources (such as Luzhkov and others in his administration) create favourable conditions for the party during elections. The ability of elites to allocate important resources to the wider public, and withhold them, and to utilise informal institutions, such as the implicit or explicit threat of dismissal for state workers, is a strong explanation for part of the party of power's electoral success in the post-Yeltsin period. The only real difference between Yeltsin and post-Yeltsin parties of power is the degree of control that incumbents (but not the party) have over political processes and the extent to which the federal executive branch is able to corral them into the United Russia organisation.

The party message and voter mobilisation

Alongside the organisational and candidate-centred explanations for United Russia's electoral success considered so far, the party message represents an equally important means of attracting voters and so of managing the outcome of elections. As noted in Chapter 2, party of power development from the early 1990s onwards saw a more nuanced approach to electoral strategy, including the use of, and even reliance upon, modern media technologies. This led one observer at the time of the 1999 State Duma election to label the party of power the 'broadcast party' (Oates 2000). In line with the discussion up to this point, the material that follows examines the contribution of United Russia's message to mobilising voters for the regime, but also the party's reliance on resources controlled by power-holders in and around the federal executive branch who remain beyond the control of any party.

Party message as a factor in United Russia's success

Party message is closely tied to party ideology, a subject discussed in more detail in Chapter 5. In general, the ideology of United Russia rests upon the three 'isms' of centrism, conservatism and Putinism, which under different guises have all figured in the party's long and difficult search for an official ideology. Centrism, in particular, has been used by party functionaries in a similar way to the 'third way' or compromise between extremes of laissez-faire capitalism and socialism (Morozov 2006). In reality, though, United Russia's centrism is far from coherent. Although conservatism and Putinism may have some broad claim to an ideological basis, centrism is better understood as a commitment to a catch-all party message that is completely in line with the evolution of the party of power electoral strategy noted earlier.

The catch-all party was first observed in post-war Germany as the *Volkspartei*, a deliberate party strategy to mitigate the tendency towards multi-partism found among the German electorate by appealing to as many voters as possible (Kvistad 1999: 65). Multi-partism in Russia, as in Germany, provides a pragmatic explanation for United Russia's adoption of a similar catch-all strategy, as, despite changes to electoral law to reduce the number of parties competing in elections, by the end of 2010 the electoral market was only just beginning to consolidate. According to a presentation given in March 2007 by the head of VTsIOM, Valerii Federov, Russian voters may be loosely grouped according to the tendencies shown in Table 3.3.[7] These tendencies are fluid, indicating that the electoral market is still

Table 3.3 Voter tendencies in the run-up to the 2007 State Duma election

Major tendencies among the electorate	Approximate percentage of the electorate	Parties competing for their vote
Communists	8–10	CPRF
Socialists – voters who do not accept bourgeois values, but who have a negative attitude to the totalitarian state. Situated to the right of the communists	4–8	A Just Russia Patriots of Russia CPRF
Social conservatives (centre-left) – supporters of a strong state and the welfare state. Strong Putin supporters	20–40	United Russia A Just Russia LDPR Patriots of Russia
Russian nationalists – this tendency is located at the periphery of the centre-left. This is the former electorate of the party Motherland. Considered too radical to be part of the communist electorate	4–9	A Just Russia LDPR
Orthodox/ monarchic – this tendency is expected to grow in the future	4–6	Congress of Russian Communities People's Will United Russia A Just Russia
Right liberals – proponents of 'social Darwinism'. They support the free market and Russia's integration with the so-called 'civilised world'	3–14	URF United Russia
Left liberals – these people reject the values of the free market and emphasise democracy and human rights as their values	5–20	A Just Russia Yabloko

Source: Round Table (2007).

very much in formation, but they do provide an overall picture of the problems of tailoring a party message to mobilise large numbers of voters. The left column shows the seven major voting tendencies evident among the Russian electorate at the time of the 2007 State Duma election, supported by core and periphery voters (centre column). The right-hand column shows the parties that, according to VTsIOM research, were expected to compete for these votes.

These voter tendencies show that the Russian electoral market does not have a large and well-defined centre as in more established democracies, where the median voter may be expected to comprise a larger proportion of the electorate. Although the centre-left represents the strongest tendency (20–40 per cent), any party message aimed at mobilising the largest number of voters must make its message as inclusive as possible to simultaneously cut across several voter tendencies. United Russia's solution to this problem has been to aim the party message at universal themes that appeal to most Russian voters, regardless of their political leanings, and to adopt a fuzzy focus on all other issues.

The 2003 State Duma election provides a good example of United Russia's catch-all strategy, understood through a systematic analysis of the party's existing programmatic material up to this point. The elections themselves were set by presidential decree for 7 December 2003, with 7 November marking the start of the pre-election campaign in the mass media, extending up to and including 5 December 2003 (Denisov 2003: 3). However, parties were mobilising and preparing for the election much earlier. For this reason the three official documents produced by the party between April 2002 and September 2003 were taken as the party's election programme as a whole.

On 29 April 2002, the party formulated its first manifesto: 'Programma Vserossiiskoi Politicheskoi Partii "Edinstvo" i "Otechestvo" Edinaya Rossiya' (Edinaya Rossiya 2002); on 23 April 2003, 'Put' Natsional'nogo Uspekha: Manifest Partii Edinaya Rossiya' (Edinaya Rossiya 2003a); and on 20 September 2003, 'Predvybornaya Programma Politicheskoi Partii Edinaya Rossiya' (Edinaya Rossiya 2003b). A sentence-level content analysis was conducted for each of these party documents, calculating the percentage of sentences devoted to (1) fact statements, (2) position issues and (3) valence issues. Position issues are those that have two clearly discernible dimensions, which divide the electorate for or against the issue in question. Valence issues, in contrast, typically involve a pair of positive or negative symbols that do not have a 'for' and 'against' component. Such issues as law and order, honesty and corruption, good times/bad times have no real alternative and provide voters with little information on the party's policy commitments, but also little reason to disagree with the party (Stokes 1992). Valence issues say nothing about how the party will achieve the outcomes it promises. By simply stating that the party wants a strong economy, wants to defend the rights of citizens, wants to eliminate corruption, etc., the voter is deprived of any real information on the position of the party. Fact statements are self-explanatory.

As shown in Figure 3.2, the number of position issues in which the party took a clear stance either for or against an issue was far outweighed by valence issues/rhetoric. In the few cases when the party did make programmatic commitments, they were in areas unlikely to alienate significant numbers of voters. For example, the party favoured the formation of a professional army, shortening the length of conscription until its realisation and raising Russia's international standing through the development of a 'mighty' military–industrial complex – issues unlikely to cause much consternation among the electorate.

Although later party documents following party congresses in April 2008 and November 2009 were to show more issue clarity than these earlier party programmes, United Russia continues in no small part to rely on non-programmatic appeals as the basis of the party message. These non-programmatic appeals show some continuity with Unity's use of symbolism in the 1999 State Duma election, including the high-profile use of celebrity, but also imagery designed to create an

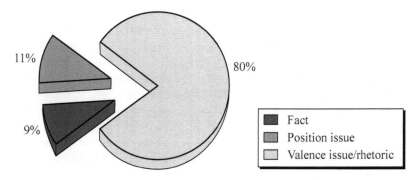

Figure 3.2 United Russia's pre-election documents (662 sentences). Source: Edinaya
Rossiya (2002, 2003a,b).

association with the CPSU. Importantly, the party continues to play on its association with the president, both Putin as former president, and Medvedev as the
current president, part of the larger locomotive strategy discussed earlier.

If in the 2003 State Duma campaign the party used the more cordial slogan,
'together with the President', by 2007 the entire election itself was transformed
into a referendum on Vladimir Putin's two terms of office. The pre-election party
document approved at the party's VIII Congress on 1 October 2007 (Edinaya
Rossiya 2007c) changed from a party programme, as was the case in 2003, to an
endorsement of 'Plan Putin', taking the title 'Plan Putina – Dostoinoe Budushchee
Velikoi Strany' ('Plan Putin – the Deserved Future of a Great Country'). The pre-
election programme wasted little time in declaring Vladimir Putin the national
leader and United Russia his political *opora* (support), despite the fact that the
approaching end of Putin's second and final term of office meant the inevitable
election of a new president during the same 2007/8 election cycle.

The same party–president association can also be seen with Dmitri Medvedev
and his 'Forward, Russia!' modernisation campaign that gained momentum from
September 2009 (Medvedev 2009). There was some speculation that the party was
on a collision course with the president's modernisation agenda when it adopted
'Russian conservatism' (socially oriented conservatism) as its official ideology at
its XI Congress in November 2009. However, a sentence-level content analysis of
the party's resultant programmatic document from the same conference (Edinaya
Rossiya 2009c) showed no contradiction. Almost a quarter of the programme (23
per cent of sentences coded) was dedicated to the goal of modernising existing
technology and infrastructure in the country.

Aside from the association with the powerful presidential figures of Putin
and Medvedev, United Russia's use of symbolism is also evident in the way the
party has targeted communist constituencies, such as pensioners and war vet-
erans, using traditional CPRF methods of electioneering, such as door-to-door
campaigning, which itself is largely symbolic in view of the party's extensive
media-based strategy. United Russia, as mentioned, has also played on association
with the CPSU, although this has, at times, been quite contradictory. The party

organisation, conferences and poster campaigns show more than a hint of Soviet nostalgia, United Russia even using the capital letter 'P' in the word 'party' in formal correspondence, just as the CPSU did in the Soviet period. However, figures within the party have (quite controversially) suggested removing a number of Soviet symbols from public life, which puts the party's attitude to the Soviet Union and CPSU in perspective.

One of the most infamous examples came in the form of a draft proposal by United Russia deputy, Aleksei Sigutkin, to remove the hammer and sickle from the Russian Army's victory banner in June 2005. This controversial issue came to a head in March 2007 when the Federation Council vetoed the draft law, which had just passed the final reading in the United Russia-dominated Fourth Duma Convocation (Gromov 2007). In the end, the problem was resolved thanks to the timely intervention of President Putin, who refused to sign the legislation into effect. In a similar way to the hammer and sickle controversy, in January 2011, United Russia deputy, Vladimir Medinskii, aired the equally controversial idea of removing Lenin's body from its mausoleum on Red Square (Medinskii 2011).

As discussed in the next chapter, the likelihood is that the party is merely testing pubic opinion on behalf of the federal executive branch regarding the removal of the last vestiges of Soviet life, as well as boosting the popularity of arbitrators such as Vladimir Putin. But in terms of its CPSU association, there is a practical logic. Aside from projecting strength and amplifying United Russia's power through association with the mighty CPSU, there is a clear electoral advantage. Pubic opinion surveys show that older voters and so those likely to carry positive estimations of the Soviet Union are among the most reliable in Russia – that is, the most likely to consistently turn out and vote in elections at every level. In the 1999 and 2003 State Duma elections, 76 and 72 per cent of older (55+) respondents said they had voted, compared with only 48 and 41 per cent of younger (18–35) respondents respectively (FOM 2003a). In July 2007, just ahead of the December State Duma election, a survey revealed age as the biggest determinant on voter turnout (FOM 2007a). Accordingly, 58 per cent of older respondents (55+) said they regularly participated in elections, followed by 43 per cent of those with higher education, 42 per cent of those without further education and 41 per cent of those living in rural areas.

Projecting success onto United Russia

There is no doubt that United Russia's own competence in pitching its party message and managing the media has contributed to the success of the post-Yeltsin regime in managing elections and gaining parliamentary majorities. However, in line with the argument so far, and in view of the prevalence of informal institutions mentioned earlier, it is important to place the communication strategy of the party in context. Russia has substantial laws governing the use of media as well as provisions to ensure fair and balanced coverage of parties during election campaigns. However, the rule of law in Russian elections, as already seen, does not always prevail.

An obvious point to mention is that United Russia's status as party agent of the federal executive branch, together with the party's recruitment of regional elites, has enhanced the party's access to national and regional media outlets, which, in turn, has led to greater bias in media coverage in favour of the party. In 2003, the ODIHR Election Observation Mission found unbalanced media coverage of the State Duma election in many of Russia's regions (ODIHR 2004). Although the ODIHR did not monitor the 2007 State Duma election, the inclusion of more governors in the ranks of the party between 2003 and 2007 is unlikely to have improved this situation. As Oleg Panfilov, head of the Centre of Extreme Journalism, notes: 'when it comes to media freedom in the regions, a great deal depends on the governors' attitude to the existence of independent mass media' (Panfilov 2007). With the process of centralisation extended to include the presidential appointment of governors from 2005, the personal attitudes of regional leaders have no doubt become much more united. Former Union of Right Forces leader, Nikita Belykh, emphasises this important point with the observation that 'Governors are always on TV during election campaigns as people who are doing their everyday jobs. So during election campaigns, United Russia has cardinal benefits in the information field' (Belykh 2007).

However, as with the primary financial flows, it is the federal executive branch that ultimately controls the primary flows of information in the post-Yeltsin period. In March 2007, the Federal Service for the Oversight of Mass Communications and for the Protection of Cultural Heritage were merged into one agency and this powerful new service is reportedly under the direct control of the prime minister, an office currently controlled by Vladimir Putin (Felshtinsky and Pribylovskii 2008: 229).

Growing state control over a variety of sources of media has already been identified as an important source of incumbent strength, one that markedly increased from Yeltsin to post-Yeltsin periods. This is particularly true for state ownership of television channels, as television remains by far the most influential source of information for voters, enhanced by the secondary nature of other forms of mass media. Surveys show that 53 per cent of Russians regularly listen to the radio (Pietilainen 2008: 369), but there are few dedicated information-based radio stations easily accessible on FM. In terms of printed press, comparative data show that Russians on average spend nine minutes per day reading newspapers (compared with an average 74 minutes for Turks), but only 6 per cent of Russians reportedly have 'trust' in newspapers (Kazantseva 2008). The internet has become an increasingly important source of information in Russia, with a reported 1,825 per cent increase in usage, 2000–10 (http:/www.internetworldstats.com), but this does not necessarily mean that Russians are using the internet as a source of political information. Research shows that the most popular entries on the Russian search engines Yandex and Rambler are for sport web pages, with only 13 per cent of Russians reading news in some form on the internet (Fossato *et al.* 2008: 14). In contrast, surveys from 2010 show that 92 per cent of respondents indicate television as their most likely source of news (VTsIOM 2010a).

Although media effects are notoriously difficult to measure, one area that is accessible to analysis is the quantity and quality of television coverage afforded to political parties. Before the 2003 State Duma election, monitoring conducted by *The Moscow Times* (15–22 September 2003) indicated that United Russia enjoyed by far the widest and most favourable coverage on all of the major Russian television channels (Dolgov 2003). Moreover, monitoring of the actual 2003 State Duma election campaign period also supports this overall conclusion, although the results were not as clear-cut as in the study by *The Moscow Times*. The ODIHR commissioned a study monitoring prime-time news broadcasts on four national television channels, 7–21 November 2003, including coverage given to every party receiving more than 1 per cent of the total airtime (ODIHR 2003). Only on Pervyi Kanal and TV-Tsentr did United Russia enjoy more coverage than any of its party competitors. On Ren-TV, for example, Union of Right Forces was the surprising beneficiary of the bulk of coverage afforded to parties, with just over 22 per cent of the overall prime-time news coverage. United Russia was some way behind, receiving just over 6 per cent. Likewise, on *NTV* these figures were 12.3 per cent and 7.1 per cent in favour of Union of Right Forces.

Although the quantity of coverage given to United Russia was by no means conclusive of systematic bias across all the channels, the ODIHR report does state that coverage given to the United Russia's major competitor, the CPRF, was mostly negative. At the same time, the ODIHR indicated that prime-time coverage of federal and regional government, as well as the president, indirectly favoured United Russia, making these figures unrepresentative of the overall media bias in favour of the party of power.

Research from the Centre for Extreme Journalism conducted between 1 October and 22 November 2007 (CEJ 2007), just prior to the State Duma election, shows a more conclusive picture of media bias. On all five national television channels, Pervyi Kanal, Rossiya, TV-Tsentr, NTV and Ren-TV, United Russia received by far the largest volume of coverage in comparison with the three other parliamentary parties that gained representation in the Fifth Duma Convocation (CPRF, A Just Russia and LDPR – see Figure 3.3). In the case of Pervyi Kanal, United Russia received nearly 14 times the amount of airtime as the next party (A Just Russia). In addition, the qualitative monitoring of these channels found United Russia's coverage to be largely favourable.

Overall, the picture is of bias, but not overwhelming bias. This is confirmed by a study conducted by Medialogia for *Nezavisimaya gazeta* for regional elections in March 2007. Although United Russia received more separate mentions (386) and more positive references (94) than any other party on the eight television channels analysed, other parties showed a surprising television presence. The CPRF received 274 mentions, 94 positive references and no negative references (Medialogia 2007).

Favourable media coverage is not always overwhelming for United Russia – coverage may change according to circumstances and the overall regime strategy towards opposition parties at any particular time. Favourable media coverage is

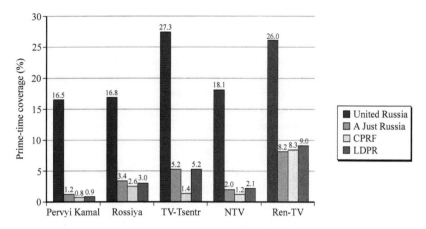

Figure 3.3 Television coverage afforded to political parties, 1 October–22 November 2007. Source: CEJ (2007).

not always a guarantee of election success either, but it does indicate the way that United Russia's favoured party agent status can translate into more and better quality airtime. Focus group work has found that most participants deny the influence of media on their choice of party, but media coverage does seem to influence certain groups of voters more than others, notably young, floating voters (Oates 2006: 129).[8] Other data also support the position that the overall effects of media may be overstated. In a survey from 2004, 43 per cent of respondents claimed that the media did not affect their voting decision, with only 9 per cent of respondents admitting that media reports clarified the electoral situation for them (Petrova 2004). However, the combination of a professional, well-thought-out electoral strategy with favourable media coverage is likely to be more advantageous than the presence of the former without the latter. In the Russian context, floating voters represent a significant portion of the Russian electorate. The head of Political Research at VTsIOM, Stepan L'vov, suggests that floating voters (*boloto*) comprise no less than 15 per cent of the Russian electorate at any one time (L'vov 2007).

It does seem likely that television exposure has allowed United Russia to build its party brand and, importantly, to establish the link between the party and the president to benefit from the strong coat-tails effect mentioned earlier. In a survey conducted two weeks into the 2003 State Duma election campaign that asked respondents to estimate the efforts of party television electioneering (Klimova 2003), 61 per cent of respondents said that they had not seen any party political broadcasts from the CPRF, with a similar 52 per cent of responses for LDPR, 61 per cent for Union of Right Forces and 63 per cent for Yabloko. Only 37 per cent of respondents replied that they had not seen any of United Russia's television agitation.

In some cases, favourable television coverage has enabled United Russia to achieve unexpected influence over potential voters without actually doing anything. In 2007, the party used changes to electoral legislation to avoid participating in televised debates during the campaign period, citing the party's desire to clarify its ideology for voters by using alternative platforms (Nagornykh 2007). In 2003, the party also declined to take part in these televised, pre-election debates, but nonetheless the bulk television exposure that the party received elsewhere contrived to create a kind of virtual participation in these debates, with survey respondents choosing the absent United Russia as the winner (Kagarlitsky 2003).

Although assessments of the positive effects of media coverage on United Russia's election results are open to question, there is little doubt as to the negative effect that this coverage has for opposition parties. The head of the Analytical Department for the consulting agency, Niccolo M, puts the challenge for opposition into perspective:

> We worked with a candidate during the 2003 [State Duma] election, with a candidate who had to fight against a United Russia candidate . . . it was very difficult because we didn't have access to media in the region. We had to come up with incredible ways to get our candidate on TV for a couple of minutes. He lost the election anyway.
>
> (Radkevich 2007)

In Medvedev's first term as president, 2008–12, there are signs that the CEC will have more power to address the problem of media coverage, perhaps indicative of the changing regime strategy towards opposition in the wake of the 2008 financial crisis. In May 2009, a law guaranteeing parliamentary parties equal access to state television and radio stations came into force, enabling the CEC to monitor the quantity of television and radio coverage per calendar month, and take measures to compensate any shortfalls (CEC 2009b). However, doubts remain over the details of the law. These doubts include the measurement of party airtime, complicated by the fact that many United Russia functionaries hold simultaneous institutional affiliations, meaning that time on television in a non-party capacity may not be taken into account. There are also concerns over its enforcement. Monitoring of five federal television channels by the CPRF's Central Committee for the month of December 2010 (CPRF 2010) showed that United Russia was afforded 19,884 seconds of airtime, followed by LDPR (3,616 seconds), CPRF (2,556 seconds) and A Just Russia (1,921 seconds), a pattern repeated in their television monitoring in previous months.

Managed elections as a feature of the post-Yeltsin period

The theoretical framework presented in Chapter 1 suggests that parties contribute to dominant-power politics by supplying consecutive election victories, which in turn serve to exclude opposition parties from the formal institutions of political power and increase the representation of regime supporters and contribute to the overall legitimacy of incumbents. There is little doubt that United Russia in the post-Yeltsin era has done this with some success. However, as discussed in the previous two chapters and as indicated in this chapter, there remains the question of the actual strength of the party as an explanatory variable.

United Russia is in many ways a net recipient of incumbent strength, and this is seen in the way that power-holders at both the regional and the federal level channel resources to the party, including favourable media coverage, candidate name recognition and the general resources offered by individual institutional affiliation. This seems to reverse the more usual principal–agent relationship found in most, but not all, political systems and resembles an earlier period in comparative politics when political parties were much weaker. In comparative perspective, the rise of political parties in the twentieth century saw the replacement of old clientelism (clientelism of the notables) with party-based patron–client relations, as parties increasingly used state resources to win electoral support (Hopkin 2001: 117). This was largely accomplished by placing party supporters in key institutional positions, giving the party control over patronage and other levers necessary to mobilise voters effectively. The twenty-first-century Russian party of power, in contrast, operates against a backdrop of a political system that was largely de-partified following the collapse of the Soviet Union. It is therefore unsurprising that the relationship between the party and its candidates and the party and power-holders in general is significantly reversed.

The effects of earlier de-partification continue to be felt even at the beginning of the third decade of post-Soviet politics. United Russia's penetration into power, as shown in the chapters that follow, is still weak. To date, parties in Russia, including United Russia, play only a peripheral role in presidential elections and presidential succession, and its powers of appointment to important institutions are limited. The material that follows examines the somewhat instrumental relationship between party of power and federal executive branch, as well as other parties in the party-system. Although United Russia is the preferred party agent of the federal executive branch and most regional administrations, it is nonetheless just one party in a multi-party strategy employed by power-holders.

Managing participation as a multi-party strategy

As noted earlier, multi-partism in the post-Yeltsin period continues to be a feature of Russian party politics, a remnant of the relatively pluralistic 1990s. From a power perspective, and in line with a theoretical framework presented in Chapter 1, it is possible to say that this legacy of multi-party politics complicates the regime's task of controlling elections through the use of just one political party. Despite United Russia majorities in both the Fourth and Fifth Duma Convocations, surveys conducted in August 2003 and August 2007 (FOM 2003b, 2007b) prior to start of each election campaign showed that the percentage of voters ready to support United Russia totalled no more than 22 per cent and 32 per cent respectively. This fact confirms the indifference of voters to party politics in general, but also the need for the federal executive branch to move beyond a one-party strategy and to manage the wider party-system to secure a constitutional majority of two-thirds of seats in the State Duma and sufficient representation in regional assemblies.

Besides the relatively low levels of support for United Russia, there are other reasons why a multi-party strategy is advantageous. Article 82 of the law 'On Elections of Deputies to the State Duma' (CEC 2005a) stipulates that at least two parties must be represented in the State Duma, accounting for no less than 60 per cent of the total vote cast. If this is not the case, then those parties that failed to pass the 7 per cent barrier must be added to the convocation in descending order until the 60 per cent minimum target is reached.

In fact, a multi-party strategy can actually work in favour of the regime and increase the prospects of Duma majorities for a single, preferred party agent such as United Russia. According to article 83 of the aforementioned law 'On Elections', the sum of votes cast for parties failing to pass the minimum threshold is subsequently redivided and redistributed to those parties that do pass the threshold. In this sense, United Russia, as the majority party in 2007, gained the most from the failure of the other seven parties to overcome the 7 per cent barrier.

By the 2007 State Duma election, and with the possible exception of Kas'yanov's and Kasparov's Other Russia movement, many of the recently formed Putin-era opposition parties had the feeling of project or designer parties, created specifically to fill a niche in the electoral market and thus support the overall objectives of the federal executive branch to ensure Duma majorities. On the left of the political spectrum, A Just Russia (created in 2006) formed part of a left-of-centre/right-of-centre tandem tactic with United Russia similar to the Ivan Rybkin Bloc/Our Home is Russia tactic of 1995. In addition, the Patriots of Russia (created 2005) also competed on the left of the political spectrum, while, on the right, Civil Force (created in 2007), a self-proclaimed representative of the Russian middle classes, appeared as an alternative to Union of Right Forces, which fell out of favour with the federal executive branch in the run-up to the 2007 election. It is interesting to note that all of these parties were reincarnations of previous parties registered with the Central Election Commission in 2002, showing the constant rebranding of 'licensed' opposition parties throughout the post-Yeltsin period. A Just Russia (registration number 5054, December 2002) inherited the registration of Motherland; Patriots of Russia (registration number 5020, April 2002) inherited the registration of the Russian Labour Party; and Civil Force (registration number 5058, September 2002) was a reincarnation of the former Russian Network Party of Small and Medium Business (see *Panorama* 2007; CEC 2008b).

To understand United Russia as part of a larger, multi-party strategy, it is important to bring into the equation some of the more established parties, as well as the more recent opposition. The legacy of multi-partism bequeathed from the 1990s left several established parties competing for a share of the vote in the post-Yeltsin period. By the end of 2010, the CPRF, LDPR and Yabloko all possess fully functioning and long-standing party infrastructures as well as recognisable party brands that offer potential problems, but also opportunities for the federal executive branch, if carefully managed. They all exist as potential components of a pro-government majority and, at the same time, their persistence helps maintain the façade of a competitive party-system. These parties provide an outlet for voters unhappy with the current political order, which is essential for maintaining political stability.

In the post-Yeltsin period, the strength of the federal executive branch provides opportunities to exert sanctions against opposition parties and even place them under its orbit of influence. This has been greatly aided by the nature of parties in Russia as top-down, leader-dominated organisations (as for Eastern Europe as a whole, van Biezen 2005: 165), making control a matter of informal agreements between party leaders and representatives of the federal executive branch. By the end of 2010, and despite frequent electoral failures and a fast-changing political environment, opposition parties show a curious continuity in their leadership. LDPR and the CPRF have thus far resisted modernising their respective leaderships, while Yabloko only replaced long-serving leader, Grigorii Yavlinskii, in 2008, despite the poor performance of the party in both the 2003 and 2007 Duma elections.

Union of Right Forces, whose party registration was officially liquated by the Justice Ministry on 30 March 2009, is a good example of the hazy opposition status of many parties. Speaking in 2007, party leader, Nikita Belykh, considered the relationship between Union of Right Forces and the authorities a difficult one, remarking that the presidential administration probably did not consider his party 'a component of their power', but that, at the same time, the party was not strictly an opposition party either (Belykh 2007). Although Union of Right Forces fell out of favour with the Putin administration in the run-up to the 2007 State Duma election, the party, as a strong supporter of the market economy, remained ideologically close to federal government. Thus, the party had communication channels with the federal executive branch through the likes of Egor Gaidar and the Institute of the Economy in Transition, and a number of past and present government ministers such as Anatolii Chubais, German Gref and Aleksei Kudrin (Gaidar 2007). In February 2009, a new liberal party, Just Cause, was officially registered incorporating Union of Right Forces, Civilian Force and the Democratic Party of Russia and including the aforementioned Kremlin insider, Anatolii Chubais, in the party Higher Council.

Without question the most significant opposition party in the post-Soviet period to date is the CPRF. The creation of the socialist-leaning A Just Russia in 2006 initially suggested that the Putin administration was looking to replace the CPRF with a controlled opposition structure. The party Motherland, which picked up 9 per cent of the vote in 2003, was part of a previous strategy to split the CPRF vote, but was an unpredictable Kremlin ally. It was incorporated into the more reliable A Just Russia in October 2006, when Motherland, the Party of Life and the Party of Pensioners agreed to merge under the leadership of Federation Council Speaker Sergei Mironov. A Just Russia clearly developed a socialist ideology to rival that of the CPRF, but tempered with a catch-all electoral strategy based around slogans such as 'new socialism' and the 'third way' (March 2009: 518), the party collectively referred to as *Esery* by segments of the media and United Russia supporters (a popular abbreviation from pre-revolution Russia, signifying members of the Social Revolutionary Party).

However, explanations for the appearance of A Just Russia are by no means clear. What is certain is that there was a great deal of intrigue surrounding the project, especially with Vladislav Surkov reviving interest in a two-party system in the summer of 2006. In August 2006, after a meeting with representatives from the Party of Life, Surkov talked of the need to create a strong challenger for United Russia for the sake of the overall health of the political system (Surkov 2006a), although there were more practical considerations involved in this party project other than the replacement of the CPRF. Chief among them was the aforementioned problem faced by the federal executive branch of incorporating independent candidates into party structures after the switch to the 100 per cent party-list format for the 2007 State Duma election. In this sense, A Just Russia also represented an attempt to integrate those elites who had previously missed their opportunity with United Russia into an approved or licensed party structure. This subject is discussed in more detail in Chapter 5.

In the final analysis, the 2007 election showed that A Just Russia failed to meet its own ambitious target of 20 per cent of the vote, and, importantly, failed to attract substantial numbers of CPRF activists as part of its attempts to undermine its main competitor for the centre-left vote. Significantly, A Just Russia also failed to gain the support of the Independent Trade Unions, an important base of support for the long-term prospects of this 'socialist' party. According to the head of the Independent Trade Unions, A Just Russia made several miscalculations in early 2007, including trying to force the Union to renege on its long-standing agreement with United Russia (Sidorov 2007).

For proponents of the so-called Mexicanisation thesis of Russian politics (Gel'man 2005, 2006; Gvosdev 2002; Riggs and Schraeder 2005), it is the CPRF at present, not A Just Russia, that plays the junior partner to United Russia, in a similar way that the Partido Acción Nacional (PAN) acted as a nominal opponent to the ruling PRI. Although the relationship between the CPRF and the federal executive branch is unclear, there is nonetheless a relationship:

> There is a special department in the presidential administration, which deals with communication with the Communist Party . . . There is a person in this department with whom we have already established . . . I would not say 'friendly relations', but normal relations, working, official relations. He even calls us and says, 'Guys I've heard that you are working on your new plan on protest actions for the next half-a-year – could you please send me a copy?'.
>
> (Savin 2007)

Even at the beginning of the post-Yeltsin period, some were noting signs of CPRF cooptation (Remington 2003). By March 2009, and in the wake of the unfolding financial crisis, information circulated that the CPRF had agreed to limit its support for protest actions in return for more media access. There were also reports that Zyuganov's own CPRF office had a direct phone line installed to communicate directly with President Medvedev and prime minister Putin (Samoilova 2009).

The Putin factor, 2000–10

As discussed in the previous chapter, the federal executive branch, notably the presidential administration, is, in formal institutional terms, the apogee of power in the Russian political system. United Russia, as an agent of the federal executive branch, is a link in the chain of command that emanates downwards from this centre (what is later termed the power vertical). As a party agent, United Russia receives support and resources that provide the party with material advantage over other party competitors. However, this support is not always material in nature. Unlike many more traditional, clientelistic parties, United Russia does not rely on dispersing financial incentives to supporters to ensure electoral success, although this chapter previously highlighted the potential of candidates affiliated with the party to do this. As noted in the next chapter, United Russia has limited control over state resources because the party has weak penetration into power. By far the biggest electoral resource for the party of power is, in fact, the popularity of the president, further reinforcing the inverse principal–agent relationship. Political analyst Sergei Markov puts this relationship into perspective: 'The purpose of this party of power is to collect popular support; if there is no popular support, they cannot collect it' (Markov 2007). The party of power collects the popularity of the president at election time; if there is no popularity to collect, as was the case throughout much of the 1990s, the entire party of power project is called into question.

For this reason it is again unsurprising that the concern with marketing and media management already discussed has not been confined to United Russia. The president is the primary beneficiary of increased media control and increasing expertise in their use – part of the process of elite learning discussed in the previous chapter. For example, a re-examination of prime-time television coverage conducted by the Centre for Extreme Journalism between 1 October and 22 November 2007 (CEJ 2007) on the five major national television channels shows that, in most cases, it was the president, not political parties, who received the majority of airtime, despite the fact that this coverage was during the Duma election campaign period. On the Pervyi Kanal, the president received 41.5 per cent of peak airtime, on Rossiya 35.9 per cent, on TV-tsentr 37.3 per cent, on NTV 46.9 per cent and on Ren-TV 13.5 per cent. On Pervyi Kanal, Rossiya and NTV the president received twice the airtime of United Russia and only on *Ren-TV* did United Russia receive more exposure than the president (26 per cent versus 13.5 per cent of peak airtime). The qualitative analysis of television coverage also conducted during this period showed that the president received far more positive coverage than any other individual or party on each of these five channels. This testifies to the continuing dominance of the president over political parties, and gives an insight as to how the patronage of a popular president has the potential to increase voter support for client-parties.

In the case of the president, it is clear that any positive association with any party is enough to influence voters and create the strong coat-tail or locomotive effect already mentioned. This explains United Russia's fixation with 'Plan Putin' in its party programme in the run-up to the December 2007 State Duma election, despite Putin's constitutional obligation to step down as president in early 2008. A nationwide survey conducted by the Levada Center between 12 and 15 October 2007, prior to the start of the official election campaign period, found that 39 per cent of would-be United Russia voters declared their support for the party because they perceived it as the party of Putin (Levada Center 2007). An open-ended survey from May 2008 that asked respondents to identify the reasons for United Russia's popularity saw the Putin factor emerge as the clear winner, named by 24 per cent of respondents (FOM 2008a).

When support for the party of power is flagging, the timely intervention of the president is enough to increase United Russia's vote share considerably. The 2007 parliamentary election provides a clear example of the locomotive strategy, with the inclusion of Vladimir Putin at the head of the United Russia party-list contributing to United Russia's overall constitutional majority. According to the head of Political Research at VTsIOM (L'vov 2007), the overnight swing of votes to United Russia after Vladimir Putin's announcement that he would head the party-list was between 6 and 7 per cent, boosting the party's rating from 47 to 54 per cent, although United Russia eventually went on to gain just over 64 per cent of the vote.

The importance of the presidential link becomes clearer when the relative popularity of the president vis-à-vis both parliament and parties is considered. Both parliament and parties continue to be viewed with suspicion by the Russian electorate. This is not a new occurrence or one that is unique to the Russian case. Rose and colleagues, writing in the 1990s, noted that, in Central and Eastern Europe, parties were the least trusted institution, with 45 per cent of citizens expressing distrust (Rose *et al*. 1998: 155). In Russia, the rise of United Russia has done little to change this fact. Research from the Russian Academy of Sciences in 2006 indicated that just over 50 per cent of respondents expressed their distrust in political parties (Osipova and Lokosova 2007: 18). In the second half of 2010, monthly surveys conducted by VTsIOM in 46 subjects of the Russian Federation showed that parties came consistently last in approval ratings for several societal institutions. In January 2011, the activities of parties received an approval rating of 22 per cent, trailing the judicial system (25 per cent) and law and enforcement officials (31 per cent), indicating the extent of public apathy (VTsIOM 2011).

In contrast, ratings for the president and (since May 2008) the prime minister continue to impress. In the case of Putin, survey data from the Levada Center show that, for almost his entire period as president of the Russian Federation (2000–8), his approval rating never dropped below 60 per cent (Levada Center 2008). Curiously, his two approval 'peaks' during this period came in December 2003 (84 per cent) and December 2007 (87 per cent) – both coinciding with State Duma elections. This may say a great deal more about the political use of opinion polling in Russia than it does about the political leadership of the country, but the popularity of figures in the federal executive branch vis-à-vis parties is nonetheless clear.

VTsIOM's 'trust index' of politicians also makes for interesting reading (VTsIOM 2010b). The index is based on monthly surveys ($n = 1,600$) that ask respondents to name the six politicians they most and least trust (open questions). This creates an index with a range of plus 100 to minus 100, with the former indicating maximum trust in the political figure in question. The first surveys in January 2006 gave Dmitrii Medvedev a rating of zero, rising to plus 34 in February 2011. Vladimir Putin's ratings January 2006–February 2011 were more consistent (plus 44 to plus 42), but the real graphic is seen in the rating of other senior United Russia figures. In the same January 2006–February 2011 period, former Unity leader and United Russia Higher Council co-chair, Sergei Shoigu, slid from plus 14 to plus 4, while Higher Council chair, Boris Gryzlov, remained around zero for the entire six-year period (plus 2 to zero).

From this perspective there are good reasons to employ a multi-party strategy at election time if there are reasons to doubt the ability of United Russia to mobilise enough support through its own popularity or the popularity of party leaders, separate from power-holders in and around the federal executive branch. Some confirmation of this point can be seen by the attitudes towards parties between election cycles. For example, research by FOM in 2006 revealed that a majority 62 per cent of respondents said that they supported no party at all, while only 15 per cent of respondents admitted to being United Russia supporters (FOM 2006). In a similar way, research by Gudkov and Dubin (2007) also puts the scale of voter apathy into perspective. More than 80 per cent of respondents questioned in February 2006 expressed an opinion that their participation in elections was unlikely to have any influence on decision-making at the state level (Gudkov and Dubin 2007: 39). Interestingly, the authors also show that support for parties, including United Russia, remains weak and indifferent. In response to the question, 'If the election for the State Duma were to take place next Sunday, would you vote, and if so, for whom?', 2002–6, the category labelled 'apathetic' received the most responses, which included those who said they would vote against all, did not know for whom they would vote, or did not know even if they would vote at all. Although the post-Yeltsin period, 2000–10, shows increasing, positive attitudes towards parties and United Russia as a longer-term trend, it would be a mistake to overestimate this trend for a party that is likely to require sustained resources, media exposure and more power (see next chapter) in order to generate genuine popular support.

Concluding remarks

In the post-Yeltsin period, 2000–10, United Russia contributed to the rise of dominant-power politics by mobilising voters to ensure election success. The State Duma has seen pro-government majorities since 2003 and party competitors continue to be marginalised from mainstream politics. However, electoral success and organisational capacity mask a more complicated reality for the party of power. Party mobilisation involves a great deal of support from extra-party resources, and even then, with the electoral arena strongly favouring United

Russia, the need for more direct intervention remains. Direct intervention can take the form of ballot stuffing or simply the growing use of locomotives, as in 2007 with the direct involvement of Vladimir Putin in United Russia's election campaign.

United Russia's role in managing elections is also shared by other parties. Aside from the proliferation of alternative parties of power in the post-Yeltsin period, the more established parties also play their part in the overall strategy of the federal executive branch. Despite the unquestionable resource advantage of United Russia, elites with access to their own resources continue to be important for continued electoral success in the regions. Although too early to talk of institutionalised support, it is clear that United Russia has succeeded in establishing only an indirect relationship with voters, mediated by the party's reliance on the media to reach voters, as well as the popularity of regional power-holders and Vladimir Putin. Although the United Russia brand is recognisable for most Russian voters, it has yet to acquire strong support among voters.

4 Governance in the post-Yeltsin period
The locale of power

Why would he [Putin] want to change the Constitution? If he introduces any changes, the parliament will be stronger, but why would he want a stronger parliament?

(Andrei Bunich 2007)

We don't have any parliamentary control; in fact we don't have parliamentary investigation. We don't approve any appointments to the executive branch, except for the prime minister, and please believe me, this is just a formality.

(Aleksandr Gurov 2007)

As shown in the previous chapter, United Russia's electoral success, regardless of the way it is achieved, has provided the post-Yeltsin party of power with the potential to influence politics at every level – from federal to municipal. In line with the theoretical framework elaborated in Chapter 1, this chapter now considers United Russia's governing role and the way that the party contributes to dominant-power politics through its activities in the legislative and executive branches as well as beyond these institutions.

Unlike the electoral (Chapter 3) and integrative (Chapter 5) roles, the role of governing and enforcing incumbent authority is, in many ways, more obvious. Comparative experience confirms that ruling parties are usually central to governance in most political systems, typically enjoying access to most, if not all, major positions of power. For this reason, this chapter is less concerned with demonstrating the strength of United Russia in this regard as it is with exploring its somewhat unusual weakness. This, in turn, is important for further developing the substantive argument of this study: that United Russia is best understood as an agent of non-party power-holders, a result of their intentions and capacities to affect larger political outcomes.

The first part of this chapter begins by establishing the legislative contribution of United Russia to dominant-power politics by examining the development of parliament–president relations in the Yeltsin period before highlighting three laws passed by the State Duma with the help of the post-Yeltsin party of power. In line with the theoretical framework presented in Chapter 1, these three laws considerably increased the position of United Russia by offering the means to control the party-system, elections and opposition forces.

In contrast, the emphasis in the remaining three sections is on the relative weakness of United Russia's governing role. The second part examines United Russia's penetration into power and what this means for the ability of the party to appoint supporters to key positions and to regulate elite competition. This includes not only the 'partification' of federal and regional governments, but also the attitude of power-holders towards the party. The third part returns to the State Duma where the contributions of the party are re-examined in order to gauge the actual strength of the party within this major party institution. Here, the theme of the elite variable is revisited from previous chapters to explain United Russia's weak governing role, but also its value as an instrument to rationalise executive–legislative and centre–periphery relations. The importance of centre–periphery relations for understanding the rationale for the all-national United Russia is then discussed as a prelude to the next chapter. The final part of this chapter examines the role of United Russia in enforcing incumbent authority outside of the formal executive and legislative institutions. This includes the party's policing function in the economy and its ability to put activists on the streets in support of the regime.

Rationalising executive–legislative relations

As already noted, political parties are considered key variables in a range of regime outcomes, although existing literature tends to emphasise the role of parties in democratic outcomes above all. Parties play an 'exceptional role' in democratic outcomes, because democratic competition ascribes them special importance in fielding candidates for elections, coordinating legislative representatives, as well as coordinating government executives (Kitschelt *et al*. 1999: 44).

However, in line with the theoretical framework presented in Chapter 1, dominant-power politics in post-Yeltsin Russia more closely resembles electoral authoritarianism than electoral democracy. Consequently, this chapter is less concerned with patterns of party–voter linkage and the quality of democratic representation, as would be the case with a normative, democratic focus. Instead, the focus is on the way that United Russia contributes to dominant-power politics, employing a power-centred rather than a normative analysis. This chapter begins by looking at changing executive–legislative relations from the Yeltsin period to the post-Yeltsin period, and the way that legislation, adopted with the aid of Unity and then United Russia, has given incumbents powerful levers to control political processes.

Opposition and government in the post-Soviet period

It is clear that the appearance of United Russia from December 2001 onwards pro-vided an important link between the federal executive and legislative branches, which in turn enhanced the ability of power-holders to propose and pass laws in a way that was not possible in the 1990s. In the Yeltsin period (1991–9), parliament proved a persistent opponent of government and a challenge to the authority of executive power in general. If the beginning of Yeltsin's presidency witnessed open conflict with the Congress of People's Deputies and Supreme Soviet during the constitutional crisis of September/October 1993, the end of the Yeltsin presidency showed that executive–legislative relations had improved little over the course of a decade.

In May 1999, the CPRF-dominated State Duma discussed the subject of impeaching the president, but the question of removing Yeltsin from power through legal means was voiced by opposition throughout 1998. In fact, 'medical impeachment' on the grounds of Yeltsin's continuing ill health was discussed by the CPRF as early as January 1997 (Rodin 1997). On 15 May 1999, Duma deputies finally voted on five charges levelled against Yeltsin, including instigating the collapse of the Soviet Union, ordering the shelling of parliament in 1993, launching the first war in Chechnya (1994–6), 'ruining' the armed forces and waging genocide against the Russian people through economic policies.[1] Although unsuccessful, the high-level discussion of impeachment nonetheless demonstrates the extent of inter-branch deadlock. Figure 4.1 provides an overview of the balance of party power in successive Duma convocations in the post-Soviet period.

The Second Duma Convocation (1996–9) was, in many ways, the epitome of inter-branch deadlock, reaffirming the belief engendered by the experience of the French Fifth Republic that the power of the president and executive branch in semi-presidential systems is only ever realised when supported by a majority in the legislative branch (Suleiman 1994: 149). As of 16 January 1996, the CPRF, Agrarian Party and People's Power deputy group, comprising largely CPRF-supporting deputies elected on the SMD portion of the vote, approached a near majority control of the State Duma, with a combined 221 seats – five seats shy of a majority (Remington 2001: 195). As detailed in Chapter 2, the poor results of the tandem tactic of Our Home is Russia and the Ivan Rybkin Bloc in the December 1995 State Duma election meant that the two most reliable sources of pro-government support in the Second Duma Convocation accounted for just 58 seats or around one-eighth of the total number of mandates.

The strength of opposition in the Second Duma Convocation meant that bargaining between the federal executive branch and party factions in the State Duma became essential for the functioning of government and the passage of legislation. To mitigate the strength of the CPRF and other opposition forces, the Yeltsin administration was often forced to rely on support from deputy groups, mainly Regions of Russia, which controlled 41 seats in the aftermath of the December 1995 State Duma election (Remington 2001: 195), but also the 51 LDPR deputies.

From the perspective of power-holders in and around the federal executive branch, neither the support of deputy groups nor the support of the LDPR faction represented an ideal situation. As Chapter 2 recounted, early experience with deputy groups elected on the SMD portion of the vote showed them to be far from reliable and, in general, they operated independently of parties, and so

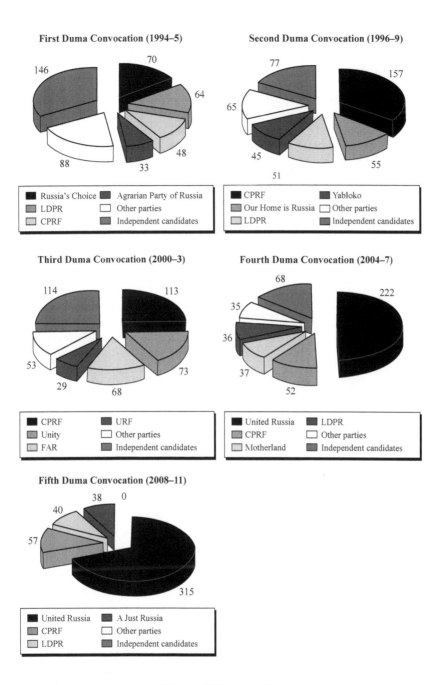

First Duma Convocation (1994–5)

146   70   64   48   33   88

Russia's Choice   Agrarian Party of Russia
LDPR   Other parties
CPRF   Independent candidates

Second Duma Convocation (1996–9)

77   157   65   45   51   55

CPRF   Yabloko
Our Home is Russia   Other parties
LDPR   Independent candidates

Third Duma Convocation (2000–3)

114   113   53   29   68   73

CPRF   URF
Unity   Other parties
FAR   Independent candidates

Fourth Duma Convocation (2004–7)

68   222   35   36   37   52

United Russia   LDPR
CPRF   Other parties
Motherland   Independent candidates

Fifth Duma Convocation (2008–11)

38   0   40   57   315

United Russia   A Just Russia
CPRF   Other parties
LDPR   Independent candidates

Figure 4.1 State Duma seats: First to Fifth Duma Convocations (largest four parties).
Source: http://www.russiavotes.org.

independently of party discipline. As for the LDPR, although its origins made this nationalistic party closer to power than the CPRF, it was by no means a reliable source of government support, even though existing literature tends to place LDPR as a party very much close to power (see Wilson 2005). Our Home is Russia deputy (1996–9) and United Russia deputy in the State Duma (2004–11), Aleksandr Tyagunov, puts the LDPR problem in perspective:

> We had contacts with Zhirinovskii, sometimes we were successful, but he's not an easy-going person. It was difficult to deal with this leader, but we were very grateful to him. Sometimes – not very often – but sometimes he supported us.
>
> (Tyagunov 2007)

Following the 1999 State Duma election, the relative success of the centrist Unity and Fatherland All-Russia (FAR) movements proved to be the beginning of a much more productive relationship between executive and legislative branches, although, as discussed in the next chapter, Unity and FAR were not identical in their positions on all issues surrounding Russia's economic and political development prior to their merger. From December 2001, the appearance of United Russia following this merger (a subject detailed in the next chapter) laid the foundation for an even more stable, pro-government platform.

One indicator of this stability can be seen in the success of the federal government in resisting opposition challenges in the form of motions of no confidence. Unlike the Second Duma Convocation, in which the CPRF narrowly missed the 300 votes needed to start impeachment proceedings against Yeltsin, the Third Duma Convocation comfortably overcame two motions of no confidence in the government. In March 2001, a CPRF motion of no confidence fell well short of the 226 or straight majority of mandates required, gaining only 127 votes in favour (Sadchikov 2001). Rather than Unity and FAR actively rallying support for the government, the motion failed largely because of the bulk of Unity, Union of Right Forces, Yabloko, LDPR and Regions of Russia deputies simply absenting from the vote (only 203 deputies were present in the voting chamber), ensuring the motion's failure (Vorob'ev 2001). In June 2003, Yabloko, with the support of the CPRF, tabled another motion of no confidence, this time gaining 173 votes in support, but still 53 short of the majority required (Chernov 2003). This last failure, just prior to the 2003/4 federal election cycle, was a timely check on opposition ambitions, preventing any early momentum they may have gained if the motion had succeeded.

In more quantitative terms, the post-Yeltsin parliament increased its law-making efficiency in direct relation to the growth of the pro-government faction in the State Duma, 2000–10. Although the subject of executive–legislative relations has received detailed coverage elsewhere in the literature (Chaisty 2005, 2008; Remington 2001, 2005) here it is illustrative to consider the quantitative output of the State Duma in the Third and Fourth Duma Convocations (2000–7) to understand the changes occurring in terms of the quantity of draft laws proposed and accepted (see Figure 4.2).

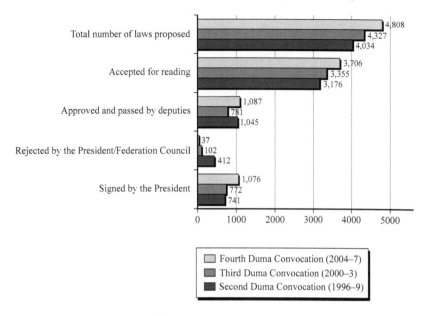

Figure 4.2 State Duma Law-Making (1996-2007). Source: Duma (2011).

As shown in Figure 4.2, the emergence of United Russia as a dominant force in the national legislature coincided with increased productivity, evident in the number of laws passed and, subsequently, rejected by the president and Federation Council. The fact that the number of laws signed by the president in this period also increased further testifies to the rationalisation occurring between executive and legislative branches as well as the personal interest shown by Putin in the legislative process. It is worth noting that, by the end of 2010, the Fifth Duma Convocation has already passed more items of legislation (1,176) that any of the four previous convocations, even though there are two more sessions left before the December 2011 State Duma election (Duma 2011).

Despite the fact that executive–legislative relations have been more productive in the post-Yeltsin period, records of the legislative activity in the Third and Fourth Duma Convocations show that the majority of laws passed had only an indirect relationship with dominant-power politics. Many of the laws passed in the period 2000–10 were designed to strengthen the effectiveness of the state and to continue Russia's consolidation of a market-based economy. These laws are too numerous to list, but in the first two years of Putin's presidency included laws aimed at the de-bureaucratisation of the economy and the reduction of administrative barriers, as well as tax, pension, banking and land reform among others (see Remington 2005: 34). For the most part, economic reforms conducted in Putin's first term of office were largely well received by international observers and liberal opposition alike.

For example, a US Congress report from 2002 noted the positive administrative and legislative changes made shortly after Putin's arrival as president, which gave Russia a more 'market friendly climate' (Goldman 2002: 21). Boris Nemtsov, a prominent opposition politician in the post-Yeltsin period, also gave a positive estimation of Putin's reforms in his first term of office noting that 'I supported Putin when he liberalised the tax-system and when he tried to protect small business from bureaucracy, when he tried to make the legal system transparent' (Nemtsov 2007). However, these positive estimations began to change during Putin's second term of office, 2004–8, as the pace of economic reform began to slow and as the political system became increasingly authoritarian.

In addition to those laws designed to increase the effectiveness of the state, there are also laws that have a more direct relationship with dominant-power politics, principally because of their obvious potential to restrict opposition forces. This is not to say that these new laws were essential for the success of power-holders in counteracting opposition forces, 2000–10. Incumbent capacity was increasing throughout the 1990s and, as Chapter 3 demonstrated, the arbitrary use of criminal law is an ever-present threat for opposition, especially when fuzzy legality encroaches on politics. Yabloko leader, Sergei Mitrokhin, provides a typical opposition perspective on this problem:

> They instigate criminal cases against activists who participate in opposition activities. These cases are completely fabricated. For example, there was a case where they accused our activist of attacking a police officer. They do this to scare young people from participating in opposition mass-actions.
>
> (Mitrokhin 2007)

Sometimes, though, United Russia has either inadvertently or intentionally helped to pass new laws that aid some of the sub-competitive aspects of Russian politics, including electoral fraud. One such example can be found in the practice of voting ahead of schedule, which was widespread in Russia between July 2005 and May 2010. This practice was made possible through a United Russia-backed amendment to an existing law (On Basic Guarantees of Electoral Rights, CEC 2002a) passed by the State Duma on 21 July 2005. The practice of voting ahead of schedule was formally designed to help overcome the logistical problems of voting in some of Russia's more remote regions, which required additional time to organise and conclude elections. Aside from this formal aspect, the localised and low-profile nature of these elections gave many informal opportunities to benefit incumbents. Ahead of schedule voting provides opportunities to hold the elections under conditions favourable for manipulation, including the use of election booths without separate cabins and elections held in administrative buildings, where the falsification of ballot papers may be easier. These elections also provide incumbents with valuable intelligence ahead of the main election and an indication as to how much falsification may be required in the election proper.

In fact, the practice of voting ahead of schedule as a tool to manipulate elections appears to be an example of an authoritarian demonstration effect or the transfer of know-how from one post-Soviet regime to another – the technique imported to Russia from Belarus, where it has featured regularly since 2001.[2] However, in what appears to be a discernible outcome of Medvedev's modernisation agenda, as well as the regime response to the 2008 financial crisis, in May 2010 the State Duma accepted the president's proposal to restrict this practice in acknowledgement of its potential for manipulation.[3]

In addition, though, many laws passed with the aid of the party of power in the post-Yeltsin period provide additional, legitimate methods for controlling opposition, reducing the need for coercion and electoral fraud to weaken opposition forces. Of all the laws passed with the support of the party of power in the period 2000–10, three in particular offer great potential for the regime to disqualify candidates, liquidate existing parties and prevent the formation of new parties and so bolter the position of incumbents. They are the law 'On Political Parties' (2001), the law 'On Opposing Extremist Activity' (2002) and the law 'On the Election of Deputies of the State Duma of the Federal Assembly of the Russian Federation' (2005). The following paragraphs focus on the details of these laws to show how they contribute to dominant-power politics.

The first of these laws was passed almost immediately after the election of Vladimir Putin as president of the Russian Federation in March 2000. On 21 June 2001, just under 55 per cent of State Duma deputies, including most Unity deputies, passed the law 'On Political Parties' at its third reading (Duma 2001). This law was specifically designed to regulate the party-system by raising the barrier for party formation and registration, while providing the state with numerous legal grounds to liquidate parties altogether. The law 'On Political Parties' currently consists of five chapters and 44 articles and has been amended on 23 separate occasions between July 2001 and November 2010.

This law is also significant in the way that it provides the first legal definition of a political party in Russia. Article 3 stipulates that parties must be all-national in scope – largely because of the requirement that a party has representation in at least half the subjects of the Russian Federation – and membership organisations – because of the requirement that they have at least 10,000 members, with at least 100 members in half of Russia's federal subjects. Importantly, the third characteristic of parties according to this law is that they compete in elections. This, in turn, means that parties are then subject to a host of further legal requirements outlined in the law 'On Elections'.

It may be argued that the Russian party-system was in need of some kind of recalibration from the Yeltsin period, but in terms of dominant-power politics the effect of this law was to reduce the space available for political entrepreneurs and the formation of new parties. Journalist Anastasiya Matveeva summaries this point clearly:

> The current party-system was formed in 2001 and at present is stable and controlled. I mention 2001 because in 2001 a new law on parties was passed meaning that the emergence of any party not under the Kremlin's control is no longer possible.
>
> (Matveeva 2007)

Table 4.1 gives a detailed breakdown of the key points of this law, showing its potential to liquidate existing parties and refuse and revoke party registration.

In fact, the use of legislation to enhance the position of insider or established parties is not confined to authoritarian political systems. Katz and Mair's elaboration of the cartel model details this point, showing how parties in democracies often use the power of the state to protect and even promote their own partisan interests (Katz and Mair 1995: 14). In a similar way, it is not unusual for consolidating democracies to legislate against perceived disruptive forces for the greater good of system stability. In post-war West Germany, for example, under the leadership of the dominant Christian Democratic Union (CDU) party, a variety of institutions and procedures were introduced to limit the threat of extremist groups, including legislation restricting regime opposition. In 1959, the Bundestag amended the German criminal code to threaten swifter punishment for treason, high treason and conspiracies against the state, although existing legislation gave plenty of grounds to imprison political opponents – in 1953 alone over 1,600 people were found guilty of 'endangering the state' (Merkl 1999: 42). This reiterates the suggestion (made in Chapter 1) that all political systems, regardless of their degree of openness and competition, face similar problems in maintaining an order or regime and, increasingly, in counteracting the threat posed by terrorism.

Table 4.1 Law 'On Political Parties' (2001): key provisions for controlling parties

Parties may be liquidated if:	• they break existing federal law (article 38) • they violate the constitution (article 38) • the party has less than 500 members in less than half the subjects of the Russian Federation (article 41) • party membership is less than 50,000 in total (article 41) • the party does not compete in elections (article 37)
Parties may be refused registration if:	• they fail to register regional branches in at least half of the subjects of the Russian Federation within six months of registration with the Central Election Commission (article 15) • they fail to pay a registration fee of 50 times the minimum monthly wage established by federal law from March of the previous year, and for the registration of regional branches a fee of three times the monthly minimum wage (article 15) • they fail to produce any of eight documents required for national registration and six documents for regional registration (articles 16 and 18) • the party undermines national security (article 9) • the party uses a name or symbol forbidden in articles 6 and 7 of the law on parties (article 20)
A party's registration may be revoked if:	• a party fails to pay a fee for modifying its statutes (three times the minimum monthly wage) (article 21) • a party fails to supply yearly information on its structure, membership and statutes (article 27) • a party does not comply with financial procedures, including supplying returns to the tax authorities every year (article 34)

Source: CEC (2001).

In the post-Yeltsin period, the national legislature passed a number of laws designed to combat the threat posed by terrorism (see Abdullaev and Saradzhyan 2006), as is the case in many countries since 2001. However, the law 'On Opposing Extremist Activity' – the second key law considered here – passed with the help of United Russia on 27 June 2002 provides significant, additional legal powers to control political opposition. This law has also been amended five times by the end of 2010. In April 2007, in an address to the Federal Assembly, Vladimir Putin spoke of the need to strengthen the law on extremism in response to the attempts of certain undisclosed elements to ignite inter-ethnic and inter-religious tensions (Putin 2007a). As a result, on 27 June 2007, a further series of amendments were passed by the Duma to improve government procedures for opposing extremism (CEC 2002b).

In fact, it has been suggested that this law, and its subsequent amendments, represents a direct replication of article 58 from Stalin's criminal code on 'counterrevolutionary agitation' and article 190 from Brezhnev's criminal code on 'knowingly distributing false information about the Soviet order' (Felshtinsky and Pribylovsky 2008: 231). For example, article 9 provides courts with the power to liquidate public and religious organisations if they are found in contradiction of the law. Article 11 deals with the mass media and extremism, requiring media outlets to self-censor, giving the courts power to confiscate materials and curtail their activities if they fail to do so. Article 14 extends the law to cover the activities of state and municipal workers, while article 16 applies to mass protests and demonstrations.

The law on extremism also complements existing laws. So, the aforementioned law 'On Political Parties' states that the activity of political parties must not breach the law on extremism (article 9) and that any party doing so may be liquidated (article 41, point 8, CEC 2001). Articles 16 and 19 of the law 'On Public Organisations' passed by the Duma in April 1995, but receiving 11 amendments between March 2002 and July 2010, excludes organisations and individuals in correspondence to the law on extremism (CEC 1995). Article 6 of the law 'On the Public Chamber' prohibits the nomination of individuals to this body from organisations convicted under the law on extremism or in receipt of a written warning regarding extremist activity (CEC 2005b).

Of course, in post-Soviet Russia, where fuzzy legality remains a problem, there is often the danger that laws can be misused or misinterpreted and applied to legitimate political opposition, rather than genuine extreme groups. In the case of the law on extremism, this problem is compounded by the interpretation of extremism, although the overall potential for misusing counter-extremism and counter-terror procedures is by no means limited to Russia. In Kazakhstan, for example, it has been noted that the sizable funding increases for the security services designed to combat terrorism have actually been used for surveillance of opposition and regime critics and for preparing cases that implicate them in criminal acts (Dave 2007: 336). The definition of extremism in Russian law is broad enough to include any proposition that threatens the multi-ethnic nature of the Russian Federation, any public justification of terrorism and terror acts, but also false public allegations against state officials. This last provision can be taken to include strong criticism against the authorities and other forms of critical election campaigning.

If the threat of this law is not enough to deter opposition, then its application demonstrates its potentially large scope. The Ministry of Justice, in accordance with article 13 of the law, publishes a federal list of materials deemed to be extremist in nature and therefore illegal. Up to March 2009, this list comprised over 360 books, articles, music albums, speeches and materials from internet sites banned in the Russian Federation, although by January 2011 this list had risen sharply to 784 items (Ministerstvo Ustitsii 2011b). Although the vast majority of these materials have a genuinely extremist content, closer inspection shows some to express little more than strong criticism of the authorities. For example, number 364 on the federal list included an article banned by the St Petersburg regional court in October 2008 titled 'Why We Will Go on the March of Dissenters' (*marsh nesoglasnykh*) on 25 November'. The Sova Centre, which monitors nationalism and xenophobia, considered this article to be one of several by the author not 'extreme' in content (Sova 2009a), although the Ministry of Justice has since removed this item from this list.

According to the Russian Ministry for Justice, by December 2010, 20 organisations have been liquidated for violating the law on extremism, including the National Bolshevik Party, banned in 2007 (Ministerstvo Ustitsii 2011c). However, the law on extremism has also been applied to organisations with a rather dubious extremist nature, including liberal-leaning Union of Right Forces in the run-up to the 2007 State Duma election, when millions of items of campaign material, from fliers to party newsletters, were seized under this law.[4] In September 2007, academic and prominent Yabloko member, Andrei Piontkovskii, went on trial for allegedly violating the law on extremism with his book *Unloved Country*, which was strongly critical of Vladimir Putin (Gomzikova and Rodin 2007).

The third important piece of legislation that provides a legal basis for controlling opposition forces is the law 'On the Election of Deputies of the State Duma of the Federal Assembly of the Russian Federation' ('On Elections') approved by the State Duma on 22 April 2005, again receiving the bulk of its support from United Russia deputies (339 deputies voting to adopt this law on its third reading, Duma 2005). This legislation, like the law 'On Political Parties', was designed to reduce the number of parties competing in elections, improving previous presidential decrees and legislation to this end.

Yeltsin's decrees for the 1993 State Duma election had some success in limiting the number of parties (in particular the signature collection requirements, see Belin and Orttung 1997: 20), but both the 1995 and 1999 State Duma elections saw a large number of parties competing. According to Golosov, a total of 35 parties collected signatures for the 1993 State Duma election, with 13 registered, but only eight crossing the 5 per cent threshold (Golosov 1999b: 108). For the 1995 State Duma election, 68 parties collected signatures, 43 were registered, but only four crossed the 5 per cent threshold (Golosov 1999b: 119). For the 1999 State Duma election, 31 associations filed nomination papers, 26 competed in the election, with only six passing the 5 per cent threshold (Rose and Munro 2002: 120). Thus, from a democratic perspective, too many parties failed to pass the threshold, leaving millions of voters without any party representation in the legislature.

The law 'On Elections' is an extremely detailed document, covering almost every aspect of elections to the State Duma, containing 14 chapters and 95 articles (CEC 2005a). The hard copy of this law from 2007 in book form is over 300 pages long and one of the frequent complaints made by opposition parties is that its complexity provides too many grounds for misinterpretation. Some of the more important provisions relevant for this discussion can be summarised as follows:

- the abolition of SMD voting in elections to the State Duma from 2007 (effectively abolishing independent candidates and consolidating parties as the main players in parliamentary elections);
- increases to the minimum threshold from 5 to 7 per cent from 2007 (effectively reducing the number of parties gaining representation in the State and regional Dumas);
- the abolition of minimum voter turnout for valid elections (a measure to insulate against threats of election boycotts);
- fixing party affiliation by banning the formation of blocs after elections within the State Duma (safeguarding against potential elite splits and shifting loyalties between elections);
- limits on campaign expenditure (ensuring that opposition parties have no financial advantage over the party of power).

In particular, the law 'On Elections' gives substantial power to election commissions to deliberate on the suitability of party candidates to compete in elections. For example, Chapter 6 of this law deals with candidate nomination and registration and totals more than 10,000 words. This process includes supplying documentation and signatures, which are then carefully scrutinised by the appropriate election commission. Yabloko leader, Sergei Mitrokhin, puts the challenge posed by this legislation in perspective:

> We were prohibited from participating in elections in those regions where it was clear that we could get a good percentage of the vote. I mean, if an opposition party, despite media and other restrictions, still manages to attract voters in some regions, the election commission would just stop them from participating.
>
> (Mitrokhin 2007)

To show the power of election commissions, in March 2007, 14 regions held elections for their regional assemblies, but only four parties were successfully registered with all 14 regional election commissions (United Russia, LDPR, CPRF and A Just Russia). The Union of Right Forces was registered in eight regions and Yabloko in only four (Moshkin and Romanov 2007). Following regional elections on the unified day of voting on 14 March 2010, the CEC received a total of 425 appeals or complaints, including 55 regarding the use of administrative resources during the elections, 75 on the conduct of pre-election campaigning and 83 on the registration of candidates. This last complaint on candidate registration was second only to the 108 complaints regarding the CEC itself (CEC 2010c).

Penetration into power

As shown in the previous section, Unity, and then United Russia, passed legislation that directly and indirectly contributed to the strength of the regime in the post-Yeltsin period. In the case of the three laws highlighted, the party of power was central in ensuring their passage through the State Duma and subsequent adoption. However, law-making is only one aspect of governing. Other aspects, such as supplying a leadership cohort and enforcing incumbent authority in wider society, depend as much on the penetration of the party into the real centres of power in the political system as a whole. This section focuses on this penetration in the post-Yeltsin period and what this reveals about the limits of United Russia's governing role.

A dominant force in the party-system or the political system?

As noted in Chapter 1, there is often little distinction made in party politics and regime studies literature between the terms 'ruling' and 'dominant' party, reflecting the fact that parties that dominate in one form or another tend to rule too. As mentioned in Chapter 3, there are exceptions, and it is by no means certain that an electorally and organisationally dominant party is also a ruling party. For the purposes of this study, a 'ruling party' is defined along the lines suggested by Geddes (2003: 52) and referred to in Chapter 1 as a party that exercises power over the leader, controls the selection of officials, organises the distribution of benefits to supporters and mobilises citizens to vote. A dominant party, in comparison, may enjoy sustained parliamentary election success over a period of time, and even organisational superiority vis-à-vis party competitors, but fail to transfer dominance in the party-system to dominance over the political system as a whole. The distinction between ruling party and dominant party is emphasised in the material that follows and revisited in the final chapter when the post-Soviet Russian party of power is considered as a type of party.

As for United Russia, the party clearly enjoys some kind of dominance, although its exact nature is often unclear or unstated. As mentioned in the introduction chapter, the rise of United Russia has been so spectacular (no less so when viewed from the weak party development of the Yeltsin period) that commentators within Russia and beyond increasingly refer to the party in terms that strongly imply a ruling party status; that is, a dominant force in the political system.

In recent years a number of articles have emerged labelling United Russia a 'dominant party' (Reuter 2010; Reuter and Remington 2009; Slider 2010) or applying the label hegemonic party to United Russia (Smyth *et al.* 2007), supported by the parallels drawn between United Russia and the 'hegemonic' PRI that dominated the Mexican political system from 1946 to 1988 (Gel'man 2005, 2006, 2008; Riggs and Schraeder 2005). United Russia deputies themselves have not refrained from drawing their own comparisons between their party and other dominant but ruling parties, such as the French Gaullist Party and the Italian Christian Democratic Party, to name but two.

The suggestion of the French Gaullist Party (Medinskii 2007) has immediate resonance, not least in the fact that the French and Russian political systems share a similar semi-presidential framework and that General de Gaulle and Colonel Putin helped to stabilise their respective political systems with the aid of their respective party support. The example of the Italian Christian Democratic Party, a party that held office for lengthy periods from 1948 to 1994, is perhaps illustrative of the historical mission of United Russia (Markov 2007), especially the way that this Italian party transformed a relatively poor country into a prosperous and economically dynamic one.

What all these party examples have in common, whether Mexican, French, Italian or similar examples, is the fact that party agents invariably controlled access to the most important power positions within their respective political systems, typically government positions, including the highest executive office, either president or prime minster. Even the commonly used hegemonic party label, according to the classic elaboration of Sartori (1976), has a clear power dimension. The hegemonic party is a ruling party that dominates power in the political system, and not just an appendage to power and an instrumental solution to the problem that elections and opposition parties pose to power-holders.

What is certain is that United Russia in the post-Yeltsin period has developed several traits associated with party dominance. First, and as mentioned in the intro- duction chapter, there is United Russia's impressive organisational complexity, extensive territorial penetration and relatively high levels of party membership. As discussed in Chapter 3, United Russia also has financial dominance vis-à-vis its competitors, and has access to resources that far outstrip those available to any other political party. In terms of candidate recruitment, United Russia's party- lists, with their large number of regional executives and institutionally affiliated candidates, certainly give the appearance of a ruling party and it is worth repeat- ing that, by the 2007 State Duma election, no fewer than 65 regional governors ran on the United Russia ticket.

Second, United Russia has developed not only physical or organisational dom- inance, but also electoral dominance. A closer examination of this electoral domi- nance through the comparative 'effective number of electoral parties' measure (Laakso and Taagepera 1979) provides a clearer picture. The number of effective electoral parties in the proportional half of the 1993, 1995 and 1999 State Duma elections (calculated according to party vote share) was 7.6, 10.2 and 6.8 respec- tively (Moser 2001, cited in Golosov 2004: 456). After the State Duma election of December 2007, the effective number of electoral parties dropped sharply to 2.2, reflecting the success of United Russia in squeezing out other parties.[5] However, the effective number of parliamentary parties in the Fifth Duma Convocation, 2007–11, calculated according to the share of seats in the State Duma, is even lower at 1.9. To put this in perspective, the effective number of parliamentary parties in the four previous State Duma Convocations was 3.7 (2004–7), 8.48 (2000–3), 6.25 (1996–9) and 9.55 (1994–5). This pattern of electoral and legislative dominance is repeated in most regions too. By 2008, the party boasted 100 per cent saturation into regional legislatures, controlling at least a third of seats in every regional Duma and over 50 per cent of seats in 36 of 83 regions (Edinaya Rossiya 2010b).

Third, there has also been a discernible increase in the formal institutional power of parties throughout the post-Yeltsin period and this is entirely in line with the comparative experience detailed in Chapter 1, with legislators, as agents of the ruling party, often (and quite rationally) consolidating their position by passing laws that favour their party. As discussed in the previous section, the amendment to the law 'On Elections' (2005) resulted in the abolition of the SMD portion of the vote in federal legislative elections, in effect signalling the end of independent, non-partisan candidates. From one perspective, this law, combined with the laws mentioned in the previous section, benefited United Russia as the 'dominant' party by removing and limiting potential and actual competition for the party. Figure 4.3 summarises the overall reduction of alternatives to United Russia over time.

Since the appearance of United Russia in December 2001, the power of the party of power has increased. On 13 October 2004, an amendment to the Yeltsin-era law 'On the Government of the Russian Federation' was passed on its third reading, receiving a total of 344 votes from United Russia, the LDPR and a number of independent candidates (Vinogradov 2004), giving ministers in the federal government the right to simultaneously hold leadership positions within political parties. At the time, this law was considered the first step in

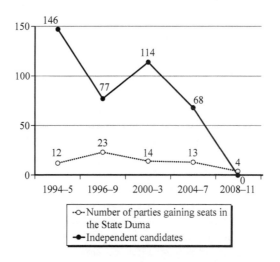

Figure 4.3 Parties and independent candidates in the State Duma, 1994–2011. Source: http://www.russiavotes.org.

establishing a legislative basis for the formation of a government from the ranks of the parliamentary majority party (Gamov 2004).

Further signs of the increasing power of parties in the post-Yeltsin period can be seen in the way that the majority party or the party holding a majority of seats in the regional legislature now has the right to nominate its own party candidate for the post of regional governor, should a vacancy arise. This change to the law, which was passed in 2008 and came into force on 1 July 2009, follows on from the abolition of direct elections for regional executives in 2005. This law allows parties holding a majority in the regional Duma in question the right to propose its candidate for the post of regional governor to the president for consideration. There is also some initial evidence that the party is having a palpable effect on the kind of governors appointed. One observer has noted that one of the features shared by many of the 27 new governors appointed by Medvedev, May 2008–10, is their 'strong ties with United Russia' (Blakkisrud 2011: 394).

Another indication of increasing party strength in the post-Yeltsin period can be found in party development outside of the legislative arena. For example, the Federation Council has witnessed a marked turnaround in the number of party-affiliated senators. Although a relatively underresearched area of Russian politics, it is easy to underestimate the 'significant constitutional power' of the Federation Council (Remington 2003: 669). The Federation Council can provide an important check on the State Duma and has the power to initiate impeachment proceedings and approve presidential decrees on the introduction of martial law, a state of emergency and the use of troops outside of the Russian Federation, among other powers (article 102, CEC 1993a).

In 2001, McFaul identified the Federation Council as a party-free state institution, reflecting the experience of the Yeltsin period, when parties struggled to attract elite support outside of the State Duma (McFaul 2001: 1163). Within a relatively short time frame, United Russia began to make inroads into this previously non-partisan institution. By 2005, of the 178 senators in the Federation Council, 102 were affiliated to United Russia and, by June 2010, United Russia listed 121 senators within the party's ranks (Edinaya Rossiya 2010c).

More significantly, by the end of 2010, United Russia also controls key leadership positions within the Federation Council. Although the prestigious Federation Council Speaker post is currently held by A Just Russia leader, Sergei Mironov, United Russia supporters control all five deputy chair posts (Svetlana Orlova, Vyacheslav Shtyrov, Aleksandr Torshin, Yurii Vorob'ev, Il'yas Umakhanov), as well as 19 of the 27 Federation Council committee chairs (Sovet Federatsii 2011). As discussed in more detail below, party supporter does not necessarily mean party member, but nonetheless it shows the palpable rise of United Russia in formal institutional terms in the post-Yeltsin period.

United Russia as a ruling party

To understand the limits of United Russia as a ruling party and by extension the limits of the party's governing role, it is important to look beyond some of these physical or cursory indicators of dominance and to consider the real penetration of the party into power. In broad terms, this can be ascertained by examining the degree to which political influence can be obtained through, as opposed to outside, party channels (Daalder 1966: 75). Specifically, the material that follows looks at the role of the party in appointing governments and regulating competition for presidential office and the overall relationship between power-holders and the party.

In fact, the question of party government is important for a number of reasons. The calculation of the 'effective number of parties', referred to above and used to indicate the dominance of United Russia, is based on the assumption that parties have government-forming potential and that for parties to be 'effective' within the political system in areas such as policy-making, appointments, patronage, etc. they have to have access to the highest executive offices. At the same time, we cannot simply assume that parties control government just because they have electoral or organisational dominance over opposition parties. Even within democratic systems, the relationship between government and the parties that support government shows a great deal of variation (Blondel 2000: 7). These relationships include the extremes of a government dominated by a supporting party to a government autonomous from party support in the areas of policy-making, appointments and patronage, among others.

Perhaps the best place to begin the discussion of United Russia's penetration into power is to examine United Russia's party leadership and the composition of the federal executive branch, including government and presidential administration. As of September 2010, almost nine years after the party's Founding Congress, the personnel that comprise the party leadership and the federal executive branch strongly diverge, with only three members of United Russia's 68-strong Higher Council represented among the major staff of the presidential administration. In government, aside from party leader and prime minister, Vladimir Putin, only three of the seven deputy prime ministers feature in United Russia's Higher Council and only three of 17 ministers.[6] In fact, the relationship between the party and federal government is at best ambiguous, and there often appears to be a game of cat and mouse played out in the media between United Russia and power-holders in and around the federal executive branch. For example, in 2005, the Finance Minister, Aleksei Kudrin, appeared in the Russian press expressing his liking for United Russia and his belief that members of government should belong to a party, but then immediately stated that he wasn't ready to join any party, but 'may be' at some point 'soon' (Kudrin 2005). By early 2010, the relationship between United Russia and Kudrin had descended into open hostility. In July 2010, United Russia deputy Andrei Isaev accused Kudrin of deliberately instigating unpopular initiatives to discredit the party, of putting a 'stick in the party's wheels' (Khamraev 2010). Perhaps as an early indication of the way the regime may sacrifice United Russia to prolong its existence, in early 2011, and in response to mass actions in parts of North Africa and the Middle East, Kudrin spoke of the necessity of 'free elections' and of introducing 'new forces into public politics' – interpreted as a direct attack on the party of power (Chernyavskii and Samarina 2011).

In a similar way, United Russia's parliamentary dominance has done little to raise the prospects of party government. Following the party's success in the December 2003 State Duma election, senior figures within United Russia, including the former Tatarstan president, Mintimer Shaimiev, called for the formation of a party-based government (Burnosov 2007), only to be rebuffed by others extolling the virtues of a professional government, that is, one not formed along party lines. Curiously, some senior leaders have actually spoken against the idea of their own party forming government. In 2006, Higher Council co-chair, Sergei Shoigu, stated his belief that government should work along professional lines and that Russia was a 'long way off' from party government (Shoigu 2006).

Despite United Russia's constitutional majority following the December 2007 State Duma election, the party failed to meet the (sometimes modest) expectations of its own leadership regarding its contribution to party government. In May 2007, Boris Gryzlov gave an interview to *Rossiiskaya gazeta* (Zakatnova 2007) stating that the next federal government must contain a percentage of ministers equal to the party's December 2007 vote share. Just before the formation of the new government, in April 2008, secretary of United Russia's General Council Presidium, Vyacheslav Volodin, gave an interview predicting that, without question, the next government 'will be a party government' (Khamraev and Barakhova 2008). However, when it came to forming this government, the man responsible – new party leader, Vladimir Putin – opted to persevere with the professional format, deciding not to elevate the power of the party at an ideal time to do so. The deputy chair of the Moscow City United Russia organisation, Andrei Metel'skii, outlines the surprising relationship between the party and federal government:

> Our government is not appointed by the party. They are not party members. The leadership of the country expresses certain sympathies to our party, but they are not our party members. In fact, they don't have much to do with the party.
>
> (Metel'skii 2007)

To date, United Russia's only foray into the realm of party government has come in the form of regional experiments, in which the party appoints its supporters to ministerial positions, but almost certainly with the prior approval of the federal executive branch. However, there is no escaping the reversed principal–agent relationship in this process. In 2006, Ul'yanovsk became the first region to have a party-based government, but this did not involve the party appointing the government in any real sense. Rather, party government happened in reverse, with the governor and nine ministers deciding to join the party en masse, with the remaining ministers joining the party later (Edinaya Rossiya 2006).

In fact, upon closer investigation this curious process of party membership after appointment, or in many cases a few weeks prior to official conferment, is a widespread practice, even when the party appears to have consolidated its position as a leading force. In February 2011, for example, four city mayors (Irkutsk, Ust'-Ilimska, Belgorod and Tambov) joined United Russia in order to strengthen the party's position ahead of the March regional elections. These mayors, with their institutional affiliation, are figures with considerable power in their respective regions, two of them gaining their positions after defeating United Russia candidates only 12 months earlier (Vezhin 2011).

In many ways, this inverse relationship between party and government is simply a continuation of the experience of the party of power in the post-Soviet period. In this respect, the party of power has always had government ministers within its ranks, but their appointment and position in government has never been dependent on the party of power. In 1999, it was EMERCOM minister, Sergei Shoigu, who formally led Unity. Likewise, for most of the Putin period, it was Minister for Internal Affairs, Boris Gryzlov, who led United Russia. Neither were agents of the party in the sense that their position depended on the continuing confidence and support of the party.

United Russia's governing role and ruling party status is also ambiguous when viewed from the perspective of the relationship between key power-holders in the Russian Federation and the party. The decision by Vladimir Putin to head the United Russia party-list in December 2007 and then to become party leader in 2008 suggests that power-holders have taken a decisive step towards institutionalising power within the party. However, Putin and his closest circle continue to have reservations concerning the party and its overall effectiveness. Putin's decision to head the party-list in 2007 may, on the surface, appear to be an endorsement of the party, but it can also be seen as an acknowledgement that United Russia had failed to generate enough popularity. By 2007, United Russia was unable to guarantee a landslide victory and constitutional majority without Putin's intervention as a locomotive, despite significant resource advantages, favourable economic conditions and opposition party weakness.

The reservations of Russia's power-holders towards United Russia can also be seen in its (continuing) instrumental use. Although playing no role of any significance in the April 2008 presidential election, United Russia was the institutional makeweight that made Putin's exit from the presidency possible. Not only did United Russia's Putin-inspired election victory in December 2007 send a very personalistic message to ordinary Russians and potential regime opponents that Putin continued to enjoy strong popular support, but the all-important constitutional majority reduced the vulnerability of Putin as he assumed the post of prime minister and party leader from April 2008. As leader of the majority party in the State Duma, Putin has some insurance against an elite schism and any attempt to remove him from this relatively weak government post.

However, the prospect of becoming a member of United Russia as well as its leader was so unappealing to Putin that party statutes had to be changed to allow a non-party member to take the post. As a party leader, but not a party member, Putin is exempt from any kind of party discipline. According to the party *Ustavy* (statutes), members of United Russia have a range of obligations, which do not apply to non-members (article 5.2). Changes were made to party statutes at the party's IX Congress in April 2008 (article 7.1.2) just prior to Putin taking up the party leader post (Edinaya Rossiya 2010d).

Putin's non-membership of United Russia also shows the danger of assuming that either high-profile party leaders or party supporters are actually party members. The law 'On the Election of Deputies of the State Duma of the Federal Assembly of the Russian Federation' (article 36, CEC 2005a) allows up to half the party-list for any legislative election to contain candidates who are not members of the party they are campaigning for. This fact alone raises questions as to the actual affiliation of many regional governors, Duma deputies and Federation Council senators.

Since April 2008, Putin has taken an active interest in party affairs, although there is a suggestion that, in order to effectively manage the party, Putin's secretariat along with trusted figures such as the mayor of Moscow, Sergei Sobyanin, are effectively running the party, bypassing the official ruling bodies. According to Gleb Pavlovskii, president of the Foundation for Effective Politics, and someone with close connections with the Kremlin, United Russia's leadership is like a group of motley trustees, who often experience problems interpreting the signals sent from the federal executive branch (Samarina and Rodin 2010), so Putin's active interest in the party may be an attempt to ensure that the signals are indeed understood, and that the party has the same objectives as the federal executive branch. As discussed in more detail in the next chapter, in particular in relation to corruption, the problem of divergent goals is a classic principal–agent problem and one that warrants extra efforts on the part of the principal to ensure that the agent is doing what it is meant to do.

In terms of party–president relations, the election of Dmitrii Medvedev in 2008 has complicated what had been a much clearer aspect of Russian politics in the post-Yeltsin period. In April 2008, Medvedev (like Kudrin) expressed his sympathy with United Russia when he spoke at the party's XI Congress, but ruled out joining the party (Medvedev 2008a). However, in his address to the Federal Assembly in November 2008 (Medvedev 2008b), Medvedev appeared to distance himself from United Russia, repeating the need to preserve the current constitution, that is, not give parliament and parties more power. By April 2009, during a meeting with senior United Russia leaders, Medvedev signalled that, despite the aforementioned changes to the law on appointing regional governors, there should be no illusion that the party would have the final decision in this key cadre question (Pozdnyakova 2009). As if to underline the limited governing role of United Russia, in the spring of 2009, Medvedev also met with opposition parties, including the CPRF, opening the way for their participation in the appointment of regional executives, with speculation that there is now a quota for opposition party governors (Rodin and Samarina 2009b).

Although by the end of 2010 there is no real evidence to support the existence of such a quota – besides the earlier appointment of former Union of Right Forces leader, Nikita Belykh, as governor of the Kirov region in December 2008 – the relationship between president and party nonetheless underlines the continuing weakness of the party. Again, at face value, Medvedev's appointment of Vyacheslav Volodin as head the government's Apparat in October 2010 may be interpreted as a presidential vote of confidence for the party, an acknowledgement of its unflinching support for Medevev and his modernisation agenda. But this development, like every other, must be placed in the context of wider elite manoeuvring that saw United Russia powerless to avert the removal of party founder and co-chair, Yuri Luzkkov, from his post as mayor of Moscow in September 2010.

The principal–agent relationship

There is no question that the presence of a majority party in the State Duma from January 2004 onwards strengthened the emerging political order. As already indicated, the law-making possibilities available to power-holders were significantly extended as the number of opposition veto players able to block the passage of legislation decreased. But in view of United Russia's weak penetration into power and the ambiguous relationship between party and power-holders, it is important to reconsider the party in a more instrumental way, as an agent of power-holders exhibiting a very circumscribed and limited governing role.

A tool for managing parliament and the party-system

As already noted, the State Duma is a major institutional arena for United Russia, one that the party dominates on nearly every quantitative indicator of significance. United Russia in the Fifth Duma Convocation controls a constitutional majority of 315 mandates, and, as in the Federation Council, oversees nearly every major leadership position. By the end of 2010, chair of United Russia's Higher Council, Boris Gryzlov, is Duma chair; Oleg Morozov, a member of United Russia's Higher Council, is the first deputy chair; and five of the eight deputy chairs are controlled by United Russia. A Just Russia (Aleksandr Babakov), the LDPR (Vladimir Zhirinovskii) and the CPRF (Ivan Mel'nikov) hold the remaining deputy chairs (Duma 2010a). United Russia also chairs 26 out of 32 Duma committees.[7]

However, there are suggestions from within the State Duma that these impressive quantitative indicators are somewhat misleading. In the post-Yeltsin period, the State Duma has emerged as a 'rubber stamp', where neither parliament nor the majority party has any real independence from the federal executive branch. In fact, as the strength of United Russia has grown, in terms of both organisational complexity and legislative representation, paradoxically its actual parliamentary role has decreased.

As shown in Figure 4.1 at the beginning of this chapter, in the First, Second and Third Duma Convocations, the balance of power was such that the federal government and presidential administration needed to negotiate and deal with a number of party factions in order to secure the passage of legislation. This was particularly true of the Third Duma Convocation, in which no one faction was able to dominate. Unity and FAR between them controlled only 141 seats, meaning that their combined pro-government support remained 85 mandates shy of a simple majority (226). This meant that negotiation was required with other party factions, in particular those government-supporting deputy groups that ran on the SMD portion of the vote, such as People's Deputies and Regions of

Russia. Former independent deputy in the State Duma, Viktor Alksnis, recalls the difference between the Third and Fourth Duma Convocations:

> At the previous Duma convocation [third] people were able to articulate opinions different from those of the executives. They were not only opposition deputies, but also members of the party of power. All of them were able to articulate their opinions even if they were against the opinions of the executive branch. All this is over now. Everything is regulated.
>
> (Alksnis 2007)

The advent of United Russia and its Duma majority reduced the need for the executive branch to negotiate with other factions and, ultimately, to consult with the Duma at all. Although the efficiency of the State Duma increased (according to the quantitative indicators presented earlier), the opportunity to debate laws before their approval has, by most insider accounts, decreased as the Duma has lost its ability to stand as an independent check on the executive branch. This situation is summed up well by a statement made by Duma chair, Boris Gryzlov, back in 2003, that the State Duma should not be an arena for political battles, but for 'effective state activity' (Vorob'ev 2003).

As an arena for effective state activity, the State Duma now deliberates less in an attempt to speed up the passage of important legislation. One of the more prevalent, informal practices from the time of the Fourth Duma Convocation onwards, coinciding with United Russia's majority after the 2003 State Duma election, is the practice of 'zero reading'. This procedure represents one of three techniques of institutional manipulation that was actually introduced in a limited fashion during the Third Duma Convocation in order to ensure the smooth passage of legislation favourable to the executive branch. Along with changes to the way that committee chairs were allocated and the creation of a coordination council to harmonise voting among pro-presidential factions, zero reading initially emerged as a means for government to reduce concessions for the annual budget by agreeing details with the coordination council before it was formally submitted to the Duma (Remington 2006: 14). Zero reading has been described as a method of adopting legislation in which the leadership of United Russia accepts bills without discussion and without the input of opposition parties, verbally agreeing any changes that need to be made within a council of experts before the bill is then introduced to the Duma (Satarov 2004).

With the advent of United Russia's majorities in the Fourth and Fifth Duma Convocations, zero reading has become less a means for ensuring the passage of legislation, as it is a technique to speed up law-making. If in the Third Duma Convocation zero reading was necessary to limit the input of the numerous opposition veto players by instigating prior discussions on draft laws between representatives of government and leaders of the four pro-presidential factions (Unity, FAR and the People's Deputy and Regions of Russia deputy groups), then from 2004 this practice largely serves to prevent the input of a large proportion of the United Russia faction, as well as limiting open debate. As United Russia deputy, Aleksei Sigudkin, notes: 'I cannot name any single important project that was not passed without the procedure of zero reading' (Sigudkin 2007). Efforts are currently under way to create similar expert councils in regional legislatures to enable zero reading beyond the State Duma (Nagornykh 2009a).

Although practices such as zero reading have supporters within the ranks of United Russia, the relentless emphasis on productivity and output is not without its critics. There is a feeling that the executive branch's desire to push through laws is sometimes counter-productive and not entirely in the longer-term interests of the state. Long-serving Moscow Legislative Assembly deputy and member of the United Russia faction, Mikhail Moskvin-Tarkhanov, comments on this subject:

> They [executive branch] tried to influence the State Duma faction and imme-
> diately got negative results – the quality of draft laws dropped sharply. The
> presidential administration was pressing the United Russia faction, telling
> deputies, 'You must do it quicker and quicker'. As a result, they adopted
> 'raw' laws.
>
> (Moskvin-Tarkhanov 2007)

There is also evidence from Duma decision-making that supports the charac-terisation of United Russia as a relatively powerless agent of the federal executive branch. If in previous Duma convocations the federal executive branch would await the outcome of parliamentary deliberation, unsure if sufficient numbers of deputies would decide to support their initiatives, then in the United Russia-dominated State Duma this situation is often reversed. There have been occasions when United Russia deputies have been left guessing the outcome of delibera-tion taking place within the federal executive branch, deliberation that ultimately determines how the party should vote on a particular issue.

A good example of this was seen early on in the party's history before the faction became more coordinated with the federal executive branch. On 12 September 2003, the People's Deputies deputy group proposed an initiative to remove the executive chair of the RAO UES energy company, Anatolii Chubais, after a series of power cuts and tariff rises questioned his ability to manage this part state-owned company. Chubais was by and large an unpopular figure anyway, while the forthcoming State Duma election scheduled for December 2003 made him an ideal target for deputies looking for a PR opportunity to increase their re-election prospects. The hope was to pass a Duma resolution (*postanovlenie*) to persuade Mikhail Kas'yanov's government, as the 52 per cent stake holder in RAO UES, to call a shareholders meeting and remove Chubais (Farizova 2003). In one day of voting, the Duma first voted to consider this initiative (306 votes in favour), but subsequently to reject the resolution (201 votes in favour), mean-ing that 105 deputies changed their minds between votes. The unlikely saviours of Chubais were the United Russia faction, who appeared to receive a counter-instruction from the presidential administration to abandon the initiative.

Likewise, there is plenty of anecdotal evidence on the internal workings of the State Duma that undermines any notion of United Russia as a genuine legislative force. For example, the decision to recall a draft law 'On Demonstrations' in 2004, according to one account, appeared to take some United Russia deputies by surprise, including the so-called authors of the law. This law had been introduced in the Third Duma Convocation in April 2003, but by April 2004 was drawing heavy criticism from segments of the media and public, especially the provisions banning demonstrations outside government buildings (Shvedov 2004). The draft law was subsequently recalled and amended in the wake of this criticism and adopted by the State Duma on 2 June 2004. A journalist who contacted one of the United Russia authors of this draft law to ask why it had been recalled:

> I called and asked him about the reasons why it [the draft law] was recalled and he said, 'but, it hasn't been recalled'. Obviously, representatives from the presidential administration hadn't called him . . . the presidential administration called him two hours after my phone call to tell him that the draft law had to be recalled. This is how things work at the Duma.
>
> (Anonymous 2007)

A tool for managing the regions

In fact, some legislation passed by United Russia undermines the very idea of a ruling party governing in its own interests. There are numerous examples of laws, either considered or actually adopted, that sat uncomfortably with party rank and file. The federal law 'On the Monetarisation of Privileges' was perhaps the best example of a law that United Russia passed despite strong opposition from elements within the party. The law was adopted by the United Russia-led State Duma in August 2004 and came into effect in January 2005, sparking weeks of street protests from pensioners and other groups across the Russian Federation. Although adopted as a federal law, Moscow City negotiated an important opt-out to maintain some subsidies and social support for certain categories of civilians, although this did not placate Moscow City United Russia deputies, who, along with many others, were unhappy that their party passed this legislation in the first place. Andrei Metelskii recalls his disappointment:

> So we adopted this law, but there were many discontented people – many. And you know; they made us to adopt this law just before elections. It was really very difficult for us to explain to people why we had adopted that law.
>
> (Metelskii 2007)

The example of the monetarisation law is revealing as it shows another instrumental aspect to the party of power – deflecting negative public opinion away from the federal executive branch. Although this law was initiated by the federal executive branch, public opinion showed that United Russia bore a disproportionate share of the popular dissatisfaction it generated. Surveys conducted by VTsIOM (2008) showed that United Russia's rating fell from 30 per cent to 24 per cent during this period (a 20 per cent drop), with levels of support for the head of state falling from 71 per cent to 65 per cent (a fall of just over 8 per cent). One report on subsequent street protests in Kavalerovo, for example, mentions that several thousand protesters congregated on the central square with banners accusing United Russia of 'betrayal' (Aleksandrov *et al.* 2005). In view of the fact that the party had limited powers to oppose this government-proposed legislation, party rank and file may be justified in feeling hard done by.

There are other examples of laws, either considered or adopted, which upon closer inspection suggest that the party is also used to gauge public opinion or even to boost the popularity of figures outside of the party. The aforementioned law on banning demonstrations had the feel of a law that was partly designed to test the reaction of the public, with United Russia the fall guy if this reaction was negative. In fact, a report appeared in the Russian press claiming on good authority that this law was indeed instigated by Kremlin staff in order to test public opinion ahead of possible tougher legislation planned for the 2007/8 election cycle (Levchenko 2007). In many ways this is unsurprising, as even in the Yeltsin period the federal executive branch used certain political parties to test ideas against public opinion, as former LDPR leader Igor Slomatin recalls: 'LDPR has been used for articulating ideas which the presidential party is not able to pronounce in public. LDPR moves certain ideas and then they check how people react to them – both in this country and abroad' (Slomatin 2007).

Sometimes laws passed by the United Russia-controlled State Duma appear to limit the power of parties per se by including provisions that restrict their scope. For example, United Russia played a key role in enabling the creation of the Public Chamber in 2005, passing the federal law 'On the Public Chamber of the Russian Federation', but as a party-free institution (article 6.2, CEC 2005b). More importantly, at least from the perspective of developing party-controlled patronage, United Russia has not used its position as a majority party to legislate greater powers of appointment, in particular in the state bureaucracy. Although comparative experience suggests that the possibilities of party-based patron–client relations are enhanced when parties predate civil service reform (Daalder 1966: 60; Epstein 1980: 104–10; Shefter 1994: 26–7), so far United Russia has not significantly influenced civil service reform in its favour.

Russian law provides the (continuing) basis for a non-partisan civil service. Article 17 of the law 'On State Civil Service of the Russian Federation' outlines several prohibitions (*zaprety*) for state servants, including their use of office to influence the outcome of elections and further the interests of political parties (article 17.12–17.14, CEC 2004). Although civil service reform is ongoing in Russia, so far United Russia has not succeeded in overcoming this remnant of earlier de-partification. According to one expert closely involved with current civil service reform, the deputy director of the Institute of State Service and Social Research, Aleksei Barabashev, the problem of the bureaucratisation of United Russia is much more actual than any partification of bureaucracy (Barabashev 2008).

This situation does have comparative parallels. Wiatr (1995) notes the same two problems in the relationship between politicians and civil servants: the politicisation of the bureaucracy and the bureaucratisation of the political system. Elsewhere the party civil service relationship has tended to vary from one political system to another. Daalder notes that, in Britain, the civil service has always remained subservient to political parties, whereas the French and German bureaucracies have dominated parties, while elsewhere the relationship has tended to be equal (Daalder 1966: 60). There is no doubt that United Russia is constantly trying to combine with components of the civil service to strengthen its position. But in line with the reversed principal–agent relationship, post-Yeltsin Russia sees the bureaucratisation of party politics, rather than the partification of bureaucracy. As such, there is a case for presenting state bureaucracy, along with power-holders in and around the federal executive branch, as another 'principal' in the post-Soviet political system in Russia.

In addition to the fact that United Russia appears to be an agent of some elements of state bureaucracy, some of the electoral reforms supported by United Russia, although helping the federal executive branch to manage and rationalise the party-system, do not necessarily benefit the party. The abolition of the SMD portion of the vote has sometimes weakened United Russia's electoral potential by curtailing the ability of the party to utilise 'local notables' running nominally independent election campaigns in regions where the party lacks popular support, but that nonetheless support the party line. In regional elections in March 2007, United Russia was sensationally beaten into second place in Stavropol'skii Krai by the newly formed A Just Russia, leaving a feeling among some United Russia supporters that changes to electoral law had not worked in the party's favour. Andrei Il'nitskii, head of United Russia Supporter Relations, observes:

> I remember how many times our party was blamed for changing election legislation – and we always told people, 'look, these changes made our party's life more complicated'. And it's true. Life showed that it's true. We would have won in Stavropol at the last elections if there were single-mandate districts. We would have won with the help of single-mandate deputies, but we lost there because the elections were only on party-lists.
>
> (Il'nitskii 2007)

These changes to electoral law, as well as the adoption of other laws detailed in this chapter, were not primarily intended to strengthen United Russia. Instead, these laws form part of a larger process of rationalising governance in the Russian Federation across several areas of the political system that were perceived as problems by the post-Yeltsin regime. This rationalisation is sometimes referred to as the creation of the power vertical, a term that can be traced to Putin's address to the Federal Assembly in 2001, when he talked of the objective of 'building an effectively functioning executive vertical power structure' to create legal discipline and an effective judicial system (Putin 2001). Essentially this involved increasing the control of the federal centre over the political system, often by simplifying the chain of command across the state, from the top down. Changes occurring in the State Duma in the post-Yeltsin period offer a perfect illustration of this rationalisation. Duma veto players were gradually reduced, but, to enhance the control of the federal executive branch over the legislative process, deputies were concentrated into one manageable structure, United Russia.

In pure management terms, United Russia was a solution to the problematic State Duma. As noted previously, parliament was by and large in opposition to government and president for much of the Yeltsin period and, by the end of the 1990s, was a power base for CPRF opposition. But even early on in the post-Yeltsin period, before the appearance of United Russia, the Third Duma Convocation was home to a variety of interests that threatened the emerging political order. Despite the presidential triumph of Putin in 2000 and the success of Unity in overcoming the FAR challenge in 1999, it was important for power-holders in and around the federal executive branch to ensure that this kind of open elite conflict did not occur again. In addition, the Third Duma Convocation was susceptible to lobbying by powerful private interests keen to buy political influence from parties and deputies.

Former Yukos owner and powerful economic elite, Mikhail Khodorkovskii, represents a prime example. Even though he initially financed United Russia, his arrest and the attack on Yukos were organised in response to his unsanctioned funding of the opposition CPRF (Overchenko 2009). Aside from using legislation and occasionally coercion to counteract this power challenge, United Russia became the tool to effectively transfer power away from the State Duma altogether. This contrasts vividly with the parties of power of the Yeltsin period that were essentially tools to transfer the Kremlin's power to the State Duma.

The main obstacle to understanding United Russia as a management tool to rationalise governance (and as a party agent in general) is that, in purely physically terms, the party appears to be very different from previous parties of power, which in turn suggests a qualitatively different kind of party, one with more importance in the political system. This image of United Russia is further reinforced by the CPSU symbolism that the party has played on, as well as by the statements of senior figures from within the presidential administration that confirm the party's higher role and historical mission. This includes the comments made by chief presidential aide, Vladislav Surkov, in February 2006 that United Russia can 'dominate the political system for the next 10 to 15 years' (Smyth *et al.* 2007: 118).

In the case of United Russia, the institutional rationale for the party of power elaborated in Chapter 2 does not fully explain the appearance of such a large, all-national party power if its only function is to supply stable State Duma majorities. From an investment perspective, the considerable resources that continue to go into building and maintaining the party seem extravagant if the organisation's only purpose is to secure Duma support. However, if we revisit a theme from Chapter 2 – that of the importance of the elite variable in the development of the party of power (intentions and strength of power-holders) – we see that United Russia was built for two purposes: to rationalise executive–legislative relations, like previous parties of power, but also to rationalise centre–periphery relations.

In the State Duma, United Russia is a solution to the problem of pluralism and competing interests, serving to control Duma deputies and limit opposition representation. Across the Russian Federation, United Russia is a solution to the problem of pluralism and competing interests by corralling elites together into one homogeneous structure, subjecting them to the constraining force of party discipline. United Russia was not created to govern, but to integrate elites across the regions into a structure controlled by the federal executive branch. This explains the physical form of the party, the weakness of the party's governing role, the strength of the party's integrative role (see Chapter 5) and the reason why the party is a dominant force in the party-system, but not a ruling party in any meaningful sense.

Throughout the post-Yeltsin period, the federal executive branch has moved decisively to alleviate a number of perceived problems inherited from the Yeltsin period that complicated governance in the Russian Federation. The most visible problem was the nature of centre–periphery relations that emerged following the break-up of the Soviet Union and which persisted throughout the Yeltsin period in the form of hyper decentralisation. As discussed in the next chapter, the Soviet-style *matreshka* federalism based along loose ethno-national boundaries provided the cleavage lines along which the Soviet Union finally collapsed. However, this process was not limited to the level of the Soviet republics, but also threatened the break-up of the Russian Federation too. In the face of a credible threat of separatism, and as a result of Yeltsin's own power struggles with Soviet president, Mikhail Gorbachev, the federal centre granted de facto autonomy to many of Russia's regions.

The actual threat of regional separatism at the time of Putin's arrival as president in 2000 is open to debate, but here it is important to note that the recalibration of centre–periphery relations in favour of the federal centre remains a key feature of the post-Yeltsin period. However, it is important to understand the exact relationship of the all-national United Russia in this process of centralisation, as it provides further confirmation of the weakness of the party as an explanatory variable in the emergence of dominant-power politics in the post-Yeltsin period.

Although United Russia helped pass legislation that directly increased the power of the federal centre over the regions, in many ways United Russia became a viable project, as well as a necessary project, only once centralisation was already well under way. As mentioned in Chapter 2, the party of power's rationale in the decentralised political system of the Yeltsin period was very different to that of United Russia in the centralised system of the post-Yeltsin period. In effect, the current party of power functions as an extra layer of administration across the regions, while providing a medium to socialise, circulate and control elites in the new political order.

The process of centralisation undoubtedly gained momentum with the creation of United Russia early on in Putin's first term of office. There is also little doubt that centralisation in the Russian Federation is an ongoing process, and so the relationship between centralisation and United Russia is one of reciprocal causation, in which the growing strength of the party of power strengthens the federal executive's control over the regions and in which control over the regions in turn opens up new opportunities for United Russia.

Although the party has helped pass numerous articles of legislation with some bearing on the centralisation process, the key ground work was already in place by the time United Russia appeared, suggesting that a United Russia-style State Duma majority was not a prerequisite for this process. Stoner-Weiss, for example, notes that the transfer of political authority from the regions to the federal centre began almost immediately after Putin's inauguration in May 2000 and included administrative changes, such as the creation of seven federal districts, the reorganisation of the Federation Council and legal measures to harmonise federal and regional law (Stoner-Weiss 2006: 148–9).

United Russia as an outcome of centralisation corresponds with implicit accounts of the party's electoral rise found elsewhere in the literature. Scholars such as Hale (2006) and Smyth (2006) put forward convincing arguments based on individual candidate strategies that centralisation, including changes to electoral legislation, limited the alternatives open to political entrepreneurs. This made running on the United Russia ticket a practical strategy for most regional elites.

The rise of United Russia also fits comparative explanations for all-national party-system development. Chhibber and Kollman (2004) argue that, in India, Britain, the United States and Canada, all-national party-systems emerged at exactly that moment when the political and economic authority of the state resided in the national government, not in the regions. In the Russian case, the increasing capacity of power-holders, in particular at the federal level, along with broad pubic and even elite support paved the way for Putin to lay the foundations for centralisation and the formation of United Russia.

In this sense, the law 'On Political Parties' adopted in 2001 became part of the early process of centralisation that prepared the ground for a much tighter, all-national party-system, dominated by one party. Political commentator, Sergei Markov, summarises the relationship between United Russia and centralisation succinctly: 'Centralisation happened first. This gave the opportunity to create a regional [party] organisation' (Markov 2007). The next chapter looks in more detail at how United Russia contributes to dominant-power politics by integrating elites across the Russian Federation.

The party as an enforcer of incumbent authority

As noted in Chapter 1, ruling parties not only contribute to dominant-power politics by passing laws in favour of power-holders; they also oversee and enforce incumbent authority in strategic areas of the state. In comparative experience, the extent to which ruling parties enforce incumbent authority, beyond the formal political institutions of power, shows a great deal of variation, but in its more extensive form it includes active involvement in managing the economy and even 'the streets' (see Chapter 1).

What is clear from the discussion in the previous sections is that United Russia's governing role, in formal institutional terms, is relatively weak and this final section provides further confirmation of this point. However, in view of the prevalence of informal institutions already mentioned in the previous chapter, some of the more subtle aspects of the structure of governance in post-Soviet Russia are considered here in order to show the difficulty in identifying the exact governing role of United Russia.

Controlling financial flows and managing the economy

Unsurprisingly, the role of United Russia in managing the economy and in overseeing the implementation of economic policy has, to date, been minimal. As discussed in this chapter, United Russia has limited control over executive decision-making at the federal level and little independence in the State Duma. This means that, even in terms of lobbying economic legislation, the party's involvement is somewhat constrained. The following comment by Andrei Bunich, director of the International Foundation 'Assistance for Business', summarises this well:

> All the decisions are made either in the Kremlin or in the government. When it comes to independent legislative initiatives in the economic sphere – well, United Russia may lobby certain decisions on selling cigarettes or beer, but nothing more significant.
>
> (Bunich 2007)

With the changes occurring between the Third and Fourth Duma Convocations, the nature of lobbying changed considerably. The so-called 'war on oligarchs' that featured in Putin's successful re-election campaign of 2004, the Yukos affair and the imprisonment of Mikhail Khodorkovskii also gave a strong message to would-be opposition to consider the risks involved in lobbying. The emasculation of the State Duma and the fact that decision-making on key issues largely occurs outside of this institution, means that the interests of big business are better served by directly lobbying federal government and the presidential administration, rather than any party faction. However, even for small and medium business, United Russia's inability to influence government and state policy has caused problems. The organisation representing Russian small and medium business, Delovaya Rossiya (Business Russia), which signed cooperation agreements with United Russia in 2005 and 2007, has hinted at the inability of the party to push the interests of the business community. In June 2009, the chair of Delovaya Rossiya, Boris Titov, questioned United Russia's lobbying effectiveness (Samarina 2009). According to Titov, even though the majority of the party faction supports tax decreases, they continue to vote for tax increases.

This is not to say that the party of power has no power to lobby on behalf of economic actors, but that, in the current political environment, lobbying is much more circumscribed and restricted. When lobbying does occur, it tends to be in areas deemed insignificant for the national interest, providing deputies with an opportunity to earn side payments, as in the Yeltsin period, but in areas deemed permissible by government or simply in areas away from the gaze of the executive branch. For example, in 2006, the State Duma ignored the concerns

of several senior government officials to adopt a series of regulations governing the sale of alcohol, termed the Unified State Automated Information System
on the Alcohol Market. Rather than strengthening the market, the adoption of
this system adversely affected the industry, and upon closer investigation the
blame fell squarely on the nexus between lobbyists and United Russia deputies.
Economist, Aleksandr Auzan, who sits on the board of the Russian Government's
Bureau of Economic Analysis, recalls the problem:

> After they started to study what happened, it turned out that these regula-
> tions were purely a Duma initiative and that the government had previously
> reacted negatively to them, but the Duma adopted them anyway. When the
> deputies were asked how it happened, they answered, 'It's not our fault. We
> never touch the voting buttons until we get the directive from the Kremlin on
> which button to press'.

(Auzan 2007)

Outside of the legislative arena the ability of the party to oversee key economic
development is hindered by the party's limited power of appointment. Here, the
presence of United Russia figures in key roles in the economy is misleading,
because these individuals owe their position to the federal executive branch, not
to the party. However, as the economy, and control over key areas of the economy,
is such an important source of incumbent strength in post-Soviet Russia, the composition of the board of directors of some of Russia's largest companies reveals
the informal structure of governance in the post-Yeltsin period, and consequently
the real relationship between power-holders and party.

As noted in Chapter 2, in the post-Yeltsin period favourable economic conditions, including increased revenue from the sale of oil and gas, were an important
factor in explaining the strength of the federal executive branch. Control over
financial flows allowed power-holders to consolidate their positions, providing
the resources to create United Russia, develop its organisation across the regions,
buy support among regional elites and gain leverage over many regions through
budget allocations. All these factors were important for the aforementioned
process of centralisation.

In reflection of the importance of financial flows for regime stability (and for
self-enrichment), key Russian companies invariably have senior representatives
from the federal executive branch overseeing their activities. Shevtsova (2007:
108), for example, lists a number of government and presidential administration
figures who served as directors for some of Russia's largest and most profitable
companies during Putin's second term as president, including the then First Deputy
Chief of Staff, Dmitrii Medvedev (chair of the board of directors of Gazprom);
First Deputy Chief of Staff, Sergei Ivanov (chair of the board of directors of
the United Aviation Building Corporation); head of the presidential administration, Sergei Sobyanin (chair of the board of directors of TVAL); deputy head of
the presidential administration, Igor Sechin (chair of the board of directors of

Rosneft'); assistant to the president, Viktor Ivanov (chair of the board of directors of Almaz-Antei and Aeroflot), among others.

It is worth noting that, by the spring of 2009, the situation had not changed greatly. Of the top ten Russian companies listed by *Ekspert* magazine in 2010, and according to the websites of the companies in question, a number of senior figures in and around the federal executive branch sat on the boards of these companies. They included Minister for Energy Sergei Shmatko (Transneft'); former presidential Chief of Staff Aleksandr Voloshin (RAO UES and Noril'sk Nikel); presidential Chief of Staff Sergei Naryshkin (Rosneft'); Deputy Prime Minister Igor' Sechin (Rosneft'); and former Finance Minister German Gref (Sberbank).

Several high-profile party leaders also serve on the board of directors of some of Russia's largest companies. By September 2010, First Deputy Prime Minister Viktor Zubkov, who previously held a leadership position in United Russia's Higher Council, sits on the board of Gazprom. Deputy Prime Minister and United Russia Higher Council leader Aleksandr Zhukov is chair of the board of directors of Russian Railways. Aleksandr Shokhin, also a member of United Russia's Higher Council, is on the board of directors of several companies, including TNK-BP, Russian Railways and Lukoil.

However, in view of the reversed principal–agent relationship and the instrumental nature of United Russia, the presence of these party functionaries does not tell us anything new about United Russia. Rather than show the strength or weakness of the party in terms of penetration into these key economic oversight positions, the composition of the boards of some of Russia's largest companies shows the persistence of what may be termed consummate insiders; that is, powerful elites in and around the federal executive branch who remain above party politics. These individuals cannot be considered agents of United Russia in any sense of the term, but form part of a power elite that resists any party control.

The flagship government initiative to create the so-called national projects is illustrative of this point. The national projects are intended to lessen Russia's dependence on oil and gas exports by channelling revenues into key strategic areas of the economy, such as investment in nanotechnology, which alone totals seven billion dollars in allocated funds. As of September 2010, this council, formed by Vladimir Putin in 2005 to oversee the realisation of these national projects, comprises 57 elites from a range of backgrounds, but all notable for their institutional affiliation in one form or another. Included in this who's who of Russia's most powerful are Vladimir Putin, Dmitrii Medvedev, Gazprom chief executive, Aleksei Miller, as well as 17 members of United Russia's Higher Council. What is noteworthy is not the number of United Russia members or leaders in this council – the party played no role in its formation. Instead, it is interesting to note the presence of LDPR leader, Vladimir Zhirinovskii, CPRF leader, Gennadii Zyuganov, and the suggestion in some quarters that this group represents an informal centre of power comparable to the politburo of the CPSU's Central Committee, that is, a ruling body with no constitutional status, but unlimited powers (Ostrovsky and Buckley 2005).[8]

As mentioned in the previous section, United Russia is something akin to a management tool, but this obviously raises questions about the identities of the managers. Throughout this chapter and previous chapters the term 'in and around the federal executive branch' has been used to signify the loose locale of Russia's highest power-holders and those who manage United Russia and reap the benefits of the party's activities. However, this appendage 'in and around' suggests that the location and concentration of actual power in the political system does not always correspond to formal office within the federal executive branch. This was certainly the case in the Yeltsin period and appears to be the case in the Putin period, indicated as much by the difficulty in claiming that the Putin period ended with election of Dmitrii Medvedev in March 2008.

When it comes to the key figures in and around the presidential administration, post-Soviet Russian experience has shown that this network of individuals can be quite broad, not necessarily corresponding with official positions. Jensen notes that Yeltsin's entourage consisted of policy experts, courtesans, family members and others, not all of whom occupied official positions (Jensen 1999: 349). Jensen goes on to note that in perhaps no other industrialised country have a car sales-man, the president's daughter, his bodyguard and his tennis coach been trusted policy advisors. In the post-Yeltsin period, the presidential entourage is a much more sober affair, but still exhibits an informal nature. The tandem or duumvirate arrangement between president and prime minster from May 2008 shows how difficult it can be for observers to identify the locale of power in Russia.

The exact nature of power in Russia is therefore open to debate and interpreta-tion. Penetrating the corridors of power is an extremely difficult task, especially the corridors of the federal executive branch (Kryshtanovskaya and White 2005). One of the more interesting attempts to focus on cliques in the Kremlin is Vladimir Pribylovskii's identification of five main groups of influence. They include the St Petersburg Jurists, the St Petersburg Chekists, the group of Nikolai Patrushev, the group of Vladimir Yakunin and the St Petersburg economists (Pribylovskii 2006). What is clear, however, is that power in Russia is not party-based, and that power-holders appoint themselves or trusted figures to manage strategic branches of the economy and the political system, including the management of United Russia.

The ambiguity of United Russia's governing role

The material presented above is completely in line with the idea of a reversed principal–agent relationship, in which a group of power-holding elites, occupying formal positions in the federal executive branch, but also other important posi-tions in the political and economic sphere (note Aleksei Miller and Gazprom as an example), control politics through a number of channels, including the party of power. In this sense, the earlier discussion on party penetration into the federal executive branch is somewhat disingenuous, because it is the federal executive branch that penetrates the party by controlling key leadership positions as well as its resources and, as demonstrated in the previous chapter, its electoral success.

However, just as power in the Russian Federation is difficult to isolate and identify, so too are the exact boundaries of United Russia's governing role. As discussed in the previous section, much depends on the elite variable; the intentions and strength of power-holders and the role they ascribe to their party agent(s). The party's governing role also depends on the opportunity structure created by fuzzy legality (see Chapter 3). This means that, in some areas, the informal governing role of the party is greater than the formal role, while in others the party seems indissolubly entwined with the state. The following examples illustrate these points.

As mentioned earlier, there are certain legal provisions that create the basis for a non-partisan civil service and this is one formal constraint on party power in general. In addition to the civil service laws already mentioned, article 10 of the law 'On Political Parties' forbids persons holding governmental or municipal office and persons in the state or municipal service from using their official position or status to promote the interests of political parties. However, the success of the party in attracting institutionally affiliated candidates makes it clear that United Russia enjoys a close relationship with many state officials and that these provisions are not always enforced. The relationship between United Russia and state bureaucracy is difficult to quantify, but it would appear that the two are interlinked in the sense that neither are subordinated to the wider interests of the public, but to the federal executive branch (although, by virtue of its expansive nature, state bureaucracy is less subordinated than the party). Here, the scope of the party governing role is greater than formal constraints suggest.

This point is illustrated well with the case of the Public Chamber, which, as mentioned, was established in April 2005 as a non-partisan oversight committee. The stated aims of the Public Chamber are to encourage cooperation between citizens, the state and local self-governance, to defend citizen rights and freedoms and to facilitate societal control over the activities of the state.

The Public Chamber, according to article 6 of the law 'On the Public Chamber' (CEC 2005b), is formed from 42 candidates, approved by the president. These candidates must not be party-affiliated, and if during the course of their work they decide to break this condition their work for the Public Chamber is curtailed. This was the case with Aleksandr Shokhin, who left the Public Chamber in 2009 to take up a leadership position with United Russia (Obshchestvennaya Palata 2010). However, it is clear that other members of the Public Chamber also have close connections with United Russia, although they may not hold formal positions in the party.

Vyacheslav Nikonov, for example, was a member of the Public Chamber until 2010, but at the same time maintains close contact with the presidential administration through his polity think tank and United Russia, participating in discussions at the Centre for Social Conservation Policy – an important United Russia think tank (see Chapter 5). Valerii Fadeev, also a member of the Public Chamber and editor of the political magazine, *Ekspert*, is director of the Institute for Public Planning and the Club 4 November think tank, an organisation with close ties to United Russia. Other figures include Gleb Pavlovskii and former State Duma deputy in the United Russia faction 2004–7, Petr Shelishch, among others. Despite the formal limits on the party's governing role, informally United Russia is more involved. A former representative of the Public Chamber who has close contacts with United Russia made the following comments:

> Look, there is a formal aspect. According to the Law on the Public Chamber, Public Chamber members are prohibited from being members of any party and/or from being directly involved in any party's activities. Therefore . . . well, it is true that I've been cooperating with United Russia very closely and this cooperation has begun in the past few years . . . but our cooperation is . . . well . . . kind of . . . is kind of consulting.
>
> (Anonymous 2007)

The same ambiguity in determining the role of United Russia can be seen in the party's activities on the streets in counteracting the threat posed by opposition activists. In Putin's second term of office (2004–8), power-holders became increasingly sensitive to the threat posed by so-called colour revolution and mass actions, in light of developments in parts of Russia's near abroad. In Serbia (2000), Georgia (2003) and Ukraine (2004), pro-democracy forces succeeded in instigating regime change, with 'youth' identified as a major factor in mobilising and uniting anti-regime forces (Kuzio 2006).

By 2004/5, the previous success of the post-Yeltsin regime in squeezing out opposition forces from the formal institutions of governance, such as national and regional legislatures, forced many parties to take to the streets as the only unrestricted arena available to them. At times, a fairly disparate array of opposition, such as Yabloko and Eduard Limonov's banned National Bolshevik Party, found common cause to protest against the emerging political order. In response, the regime counter-mobilised to restrict this space and to consolidate the image of grass-roots regime support for domestic and international media.

In fact, from 2005, there was a discernible proliferation of youth movements, in particular pro-regime movements – a clear sign of the regime's strategy to prevent a Ukrainian scenario from unfolding in Russia. In line with comparative experience detailed in Chapter 1, some of these pro-regime youth organisations began to perform something akin to a policing function, supplementing the work of existing law enforcement agencies. For example, reports appeared in a local Moscow newspaper in 2007 (Trekhov 2007), claiming that the pro-Kremlin youth organisation, Nashi, had struck an agreement with the authorities to 'patrol the streets', while another pro-regime youth organisation, Mestnye, had already completed a 'raid' with the Federal Migration Service, helping to detain 72 illegal immigrants.

However, the role of United Russia in deploying these or other youth organisations is difficult to gauge, owing to the fact that many United Russia deputies are unaware of the youth groups actually affiliated with their own party. According to a figure in the Club 4 November think tank, activists from the youth movement Nashi attended courses run by this organisation (Mekhanik 2007), but several senior figures in United Russia seemed unsure if this particular youth organisation had any formal connections with the party.

Again, at face value, United Russia seems well equipped with its large organisational network and membership to deploy supporters on the streets at short notice. In January 2009, the party reportedly mobilised 10,000 supporters in Khakasiya to back President Medvedev's anti-financial crisis measures (VG-news 2009), while May Day and Victory Day celebrations usually see some demonstration of party support. But there are grounds to question the actual potential of the party to deploy supporters on the streets and to ask if it is the party that mobilises supporters or the federal executive branch, regional administrations and the state apparatus, using the party as the shell to create the appearance of societal support. For example, Molodaya Gvardiya, a youth organisation considered to be United Russia's own party youth wing and one active on the streets in support of various regime aims, apparently receives no funding at all from the party, while there is some uncertainty surrounding the subordination of this youth wing to the party central office and leadership. Andrei Il'nitskii in United Russia's Central Executive Committee comments:

> we don't restrict this organisation by imposing our strict management requests and we do not influence their activities by financing them directly or indirectly. No – they work independently – I don't know the sources of their financing. I guess these sources are youth foundations.
>
> (Il'nitskii 2007)

While the statutes of this youth organisation give little information on the sources of its funding, aside from citing membership contributions, donations and revenue from events (Molodaya Gvardiya 2009a), this official party statement appears to contradict comments made by a senior figure within Molodaya Gvardiya on the same subject. Nadezhda Orlova, former chair of the Molodaya Gvardiya Federal Political Council, 2008–8, and as of March 2011 a United Russia State Duma deputy (replacing Valerii Komissarov who resigned his seat), makes the following comment:

> Our relations are based on the principle of equality, but the party finances us. I think you might be interested in financing in general – so I will tell you before you ask: the scheme of financing is different for every region, but on the whole, all the money, as a rule, comes from the party.
>
> (Orlova 2007)

Understanding United Russia's governing role, especially the enforcement of incumbent authority on the streets or in other areas, is difficult to gauge, because it is difficult to distinguish between formal and informal power and to clearly separate the party from the state. This situation perhaps explains why it is difficult for many United Russia deputies and rank and file members not only to identify their own youth movements, but also to put the achievements of the party in the Putin period into concrete terms. Andrei Metel'skii, speaking of the achievements of United Russia, seems to make no distinction between the achievements of the Moscow City Duma faction, the achievements of the Moscow City administration and the wider functioning of the state:

> Our faction in the Moscow City Duma makes up the majority – there are 28 United Russia deputies out of 35. Naturally, no law is possible to adopt without us . . . look around, you can see that the city has electricity, public transport, new roads are being constructed . . . kids go to school, to kindergartens, teachers are paid their salaries – and all this is the result of our work.
>
> (Metel'skii 2007)

The party slogan referred to in Chapter 3, *partiya real'nykh del* (the party of real achievements), seems equally ambiguous when the party is run as an extension of the state, but has so little independent decision-making power. There is no question that the party has a large budget to launch its numerous party projects and to create a very visible presence, giving the impression that the party is busy running the country. However, it would be a mistake to overestimate the role of the party, even when it comes to administering its own party projects. Vadim Bondar', a former Union of Right Forces State Duma Deputy (2000–3) and someone who works extensively in the sphere of local self-governance, recalls his own personal story on this subject:

> I had an assistant. She worked for an association of parents with disabled children. There were 600 families in this association. They were poor people and weren't able to pursue their projects to help kids, but the organization was growing and in the end representatives from United Russia came to them and offered money for a joint project. My assistant said, 'The money was good, the project was good – all we needed to do in return was just to say that it was United Russia's project and that it was their "real" achievement.'
>
> (Bondar' 2007)

This raises an important question on the nature of United Russia's contribution to governing the Russian Federation, but more importantly the relationship between the party and regional administrations and with society in general. Overall, the party lacks any real independence and, up to the end of 2010, has little value in and of itself, beyond its value as one of many organisational tools at the disposal of power-holders. The next chapter considers United Russia's links with society and the formal structure of the party, including its relationship with regional administrations across the Russian Federation.

Concluding remarks

United Russia has clearly contributed to dominant-power politics in the post-Yeltsin period by providing the mechanism for rationalising executive–legislative relations, and, as shown in the next chapter, rationalising centre–periphery relations too. What this chapter has shown is that United Russia's governing role is much weaker than the formal appearance of the party suggests. In many ways, the actual role of the party of power has paradoxically decreased over time, as independent sources of power, such as the State Duma, have receded.

The key point is that the principal–agent relationship is significantly reversed. Power-holders continue to operate beyond the control of any party, meaning that United Russia, and the party-system in general, is simply managed by the federal executive branch as another strategic area of the state. The next chapter extends the discussion from the end of this chapter to consider United Russia's links with society, its formal party organisation and its overall integrative role. This is crucial for understanding the post-Yeltsin party of power as a party that was created not to govern, but to manage the party-system, manage executive–legislative relations, but also centre–periphery relations – to integrate elites from across the regions into a structure controlled by the federal executive branch.

5 The politics of stability

United Russia as an integrative force

> What United Russia does is shown by its name. Its name reflects the fears of those
> days – in the period from 1999–2003; the fears of territorial disintegration. Yet,
> these fears haven't gone away.
>
> (Andrei Chadaev 2007)

> I don't think that the power vertical is very effective and it's getting less and less
> effective.
>
> (Georgii Satarov 2007)

As suggested in the previous chapters, a large part of the rationale for the all-
national United Russia was the process of centralisation and the belief that a party
organisation could provide an important, extra layer of administration across the
Russian regions. Although the party's role in governing the Russian Federation
is relatively weak and although the party's role in managing elections is, so far,
reliant on non-party resources, such as state media and presidential popularity,
there are reasons to expect a greater integrative role for the party.

The first part of this chapter outlines the reasons for this expectation as well
as the larger rationale for United Russia. This includes analysis of structural factors that made and make an all-national party desirable, including Russia's model
of federalism as it emerged after the collapse of the Soviet Union. In line with
the importance of the elite variable identified in Chapter 2, the first part of this
chapter also considers the influence of agency in the creation of United Russia.
This includes the 'Putin factor' and the influence of his leadership style on the
development of the party of power.

This chapter then shifts attention to the potential of United Russia as an integrative force, detailing the formal organisation, its physical characteristics, links
with other organisations and capacity to control and socialise party members.
The third part assesses the ability of the party to attract supporters through the
combination of selective and collective benefits. Finally, this chapter strikes some
balance by considering the limits of United Russia's integrative potential. The
top-down origins of the party of power lend it more to elite integration than
to the integration of society as a whole. At the same time, the way that United
Russia penetrated the regions means that, as well as a party agent for the federal

executive branch, in some regions the party has become an agent for regional administrations. This questions the value of the party as a tool to keep regional administrations in check.

The integrative rationale of an all-national party

In October 2007, during a live phone-in with members of the public, President Putin acknowledged the importance of United Russia in the country's post-Yeltsin development, referring to the party as a 'political force' (Putin 2007b). What is clear from the discussion so far is that the basis of this assessment is not quite so straightforward and that United Russia is not the force we are led to believe. Although previous chapters show the party's positive contribution to the regime, in terms of collecting votes and rationalising executive–legislative relations – factors in the persistence of dominant-power politics – the party is still very reliant on extra-party resources and (so far) has little independence from its executive branch sponsor. At the same time, neither United Russia nor its predecessor, Unity, can be said to be causal in the appearance of either Putin or Medvedev as major political actors, while the fate of the country's top leadership remains beyond the control of any party. In line with the argument developed so far, United Russia reflects the elite variable, or the intentions and ability of power-holders to affect outcomes and to project their power onto party agents.

However, as the previous chapter indicated, and as stated at the beginning of this chapter, there are a number of reasons to expect a stronger integrative role for the party and so a clearer relationship between the party and the overall stability of the post-Yeltsin order. The following material considers dominant-power politics from the perspective of party integration, as well as the larger rationale for the post-Yeltsin party of power in more detail.

United Russia and the regional connection

As mentioned in the previous chapter, the fact that United Russia has a very different physical form from its Yeltsin-era predecessors weakens the thesis proposed in Chapter 2 that the party of power is an incumbent response to the constraints of semi-presidentialism. According to this thesis, institutional arrangements make the full realisation of presidential power reliant on pro-government support in the legislature in order to reduce the prospect of inter-branch deadlock. Consequently, parties of power are designed to provide this support, and to lessen the influence of opposition veto players over government policy initiatives. The logic of this argument is called into question with the appearance of the complex, all-national United Russia, if its overriding goal is simply to provide stable majorities in the national legislature.

To make more sense of United Russia's rationale, it is important to look beyond semi-presidentialism and to consider the overall development of the Russian state as it emerged from the Soviet Union in 1991. The balance of institutional power, in particular the loose system of federalism that saw the growing autonomy of regional administrations from Moscow, affected the calculations of the authors of United Russia just as much as the necessity of pro-government support in the State Duma.

In fact, centre–periphery relations have always influenced the modus operandi of the party of power. As noted in Chapter 3, Unity's 1999 election campaign carefully created the image of a 'party of the regions', exploiting existing voter distrust of Moscow, allowing the party to avoid negative associations with the old Moscow-based Yeltsin order. Prior to the appearance of Unity, the structure of Russian federalism afforded parties of power a weak, but by no means inconsequential, linkage function between Moscow and the regions. It has been noted elsewhere that Our Home is Russia provided (additional) access to the federal executive branch for regional executives while at the same time enjoying some success in organising support for government polices among the regions in the Federation Council (Oversloot and Verheul 2000: 127). In a similar way, the Democratic Russia movement that formed a key component of the Russia's Choice bloc in the 1993 State Duma election also functioned as the 'eyes and ears' of federal government, providing intelligence on the privatisation process throughout the regions. The Public Committee on Russian Reforms, charged with the task of monitoring privatisation, was largely built on the base of Democratic Russia's regional organisation. Mikhail Shneider, a prominent Democratic Russia and Russia's Choice activist, recalls their input: 'We provided information on how things were going in the regions in relation to Gaidar's declarations and government directives. We provided information on how all the directives and ideas were being implemented at grass-root level' (Snieder 2007).

In the case of United Russia, the situation is somewhat different. In contrast to the Yeltsin-era parties of power, United Russia was determined in no small part by the (ongoing) centralisation process. To summarise the argument already presented, in the post-Yeltsin period, power-holders in and around the federal executive had both the intention and the ability to centralise power and this, in turn, made an all-national party of power both possible and desirable.

At face value, the way that Russia emerged from the Soviet Union in 1991 created certain tensions between centre and periphery as well as the basis for multiple, and even competing, centres of power within one state. Out of the 15 Soviet Republics that emerged as independent states following the collapse of the Soviet Union, only Russia was subsequently (re)established as a federation, with the remaining 14 emerging as unitary states.[1] Russia, as a huge, multi-ethnic territory, is exposed to unique pressures, and, as discussed below, certain centrifugal forces.

Interestingly, the idea of creating more control for the federal centre over the regions, later termed the power vertical, was one of several ideas prominent in the post-Yeltsin period before the arrival of Vladimir Putin as president in 2000. This, in turn, suggests that the post-Yeltsin period simply signifies the ascendency of an older, statist tendency within the post-Soviet Russian, and even Soviet, elite. Just as Chapter 2 identified the Soviet origins of the post-Soviet party of power, the post-Yeltsin power vertical was formally voiced in the Yeltsin period by former prime minister and one-time would-be presidential opponent of Putin, Evgenii

Primakov. In St Petersburg in February 1999, Primakov stated that in the next year it would be necessary to amend the constitution to establish a 'strict vertical' between the federal centre, governors and municipal organs of power. Primakov also suggested that a new system of elections for governors was required, one in which the president would play a greater role (Slyusarenko 1999). This latter point clearly predates Putin's abolition of gubernatorial elections in 2005 by some way.

Regardless of the origins of either the power vertical or this ascendant statist element, by 2000 the decentralised politics of the Yeltsin period were considered a problem in need of urgent resolution. This problem was and continues to be framed in terms of a 'threat', one that highlights the real possibility that the Russian Federation may follow the Soviet example and split along ethno-national lines – most likely following existing territorial boundaries. From the very outset of the post-Soviet period, observers have noted a number of potential and actual separatist claims in the regions of Russia in support of this hypothesis. In the early 1990s, Chechnya became the obvious focus of attention, but the threat of separatism was also evident in other national republics and even some ethnically Russian regions too. Moser (1998), for example, documented the rise of regionalism in Sverdlovsk and the idea of a Urals Republic that surfaced at the time of the 1993 State Duma election. Former Yabloko deputy, Petr Shelishch, who joined the United Russia faction in the Fourth Duma Convocation, recalls a similar problem in St Petersburg in the 1990s, which persuaded him that the threat of separatism was real:

> I will give you an example, which is convincing for me . . . in St. Petersburg an organisation was created called 'the Society for the Separation of St. Petersburg from Russia'. Their logic was understandable – there was a feeling that the ship was sinking.
>
> (Shelishch 2007)

At the very end of the 1990s and into the post-Yeltsin period, the potential for regional separatism remained palpable. Chechen incursions into neighbouring Dagestan in 1999, followed by the Second Chechen War, underscored the continuing problem in this particular national republic, but in other national republics there were signs of tension with Moscow too. In Bashkortostan, for example, there has been an ongoing debate on the nature of the republic's identity dating back to the Yeltsin period (Graney 1997). This has mainly centred on the key issue of the status of Russian as the official language of the republic, exacerbated by Bashkir nationalists who consider themselves the only true indigenous people of the region.

However, a number of demographic considerations make Russia less vulnerable to fragmentation along ethnic lines than, say, Yugoslavia or the former USSR. In demographic terms, Russia is quite homogeneous, despite the presence of more than 160 nationalities. According to the 2002 Russian census, seven nationalities have populations of more than one million (Tatars, Ukrainians, Bashkirs, Chuvashes, Chechens, Armenians and Russians), but Russians make up the largest, totalling 116 million, around 80 per cent of the population (Perepis' 2002). Although evidence for 'ethnification' or nationalising policies in republics such as Bashkortostan should not be underestimated, they must be placed in context. Cleavage lines are not always drawn against Russians. It has been noted, for example, that in Bashkortostan the Bashkirs are more concerned about the relatively large Tatar minority than about ethnic rivalry with Russians (Hughes 2001: 134).

So, gauging the overall threat of separatism for the territorial integrity of Russia is a difficult task and one that is certainly beyond the scope of this study. Instead, the threat of separatism should be seen in terms of a powerful discourse in the post-Soviet period and one that remains firmly embedded within the Russian psyche. This is true at the level not only of power-holding elites, but also of ordinary Russians. According to a survey conducted in 2001, 44 per cent of Russians saw other nationalities (not only Chechens) as their biggest security threat (Rose 2001: 52). Although there is no analogous opinion poll data for 2010, it is interesting to note that Russian surveys show a fairly consistent quarter of respondents agreeing that the aim of terrorism and terrorists is in some way to 'take territory' away from Russia (VTsIOM 2010c). Despite the centralisation of the post-Yeltsin period, the threat of state break-up is very much ongoing too. The American National Intelligence Council's 'Global Trends 2025' implicitly highlights demographic issues in Russia and in other regions that may lead to future separatist claims (NIC 2008: 24–5), although this was not as stark as the previous 'Global Trends 2015', which carried an explicit and even imminent warning of Russian state breakup (see Osborn 2004). Opposition figures (Ryzhkov 2011) continue to claim that too much centralisation, rather than too little, currently threatens Russia's territorial integrity.

At a discursive level, the threat of separatism and state break-up provides a ready-made rationale for centralisation and for United Russia, but at a practical level so too do the difficulties of governing a fragmented state. This practical problem (rather than discursive threat) represented a real challenge to the authority of the federal executive branch and those who desired a strong state rather than strong regions. The asymmetrical federalism that emerged in the Yeltsin period, with bilateral deals brokered between the federal centre and individual regions, arguably reduced the prospects for any coordinated, political and economic reform of the country. In short, the loose federalism of the Yeltsin period tolerated legal inequalities between regions, while the ability of regional leaders to build successful political careers, independent of Moscow, reduced their incentives for obeying directives from the federal centre.

To illustrate this latter point – the relative autonomy enjoyed by many regional leaders – it is worth considering the basis of political power in the regions in terms of the aforementioned structure of governance. As already noted in previous chapters, the ability to utilise the resources of institutional affiliation (Golosov 2004), the persistence of patrimonial communism (Hale 2003; Kitschelt 1995: 453; Lynch 2005: 128–65) and the strength of informal institutions create a structure of governance altogether favourable for incumbents. This structure of governance enhances the strength of federal-level power-holders, but so too power-holders at the regional and municipal levels.

An example of this can be seen in the success of many regional heads in perpetuating their political careers through the ballot box, in particular in the period up to 2005 when regional executives were popularly elected, not appointed by Moscow. The ability of regional elites to utilise available resources and exploit local conditions has been well documented in the literature (Golosov 2004, Hale 2006; Smyth 2006). Golosov (2004), in what is probably the most detailed investigation into regional elections in Russia to date, found that, during the second election cycle, 1995–9, incumbents prevailed in 51.1 per cent of executive elections in the regions. In the third election cycle, 1999–2003, this figure was even higher at 63.2 per cent (Golosov 2004: 143). Although these figures may not be particularly striking, especially the re-election rates from the second election cycle, they are more convincing when considered in context. The 34.1 per cent of successful challengers during regional executive elections, 1995–9, were also institutionally affiliated, occupying positions of power within regional administrations as assembly leaders, city leaders, state and municipal officials or Duma deputies (Golosov 2004: 143).

The process of elite learning, mentioned in Chapter 2, was integral to this growing independence, as regional elites became more competent in managing elections and utilising resources, but their relative political autonomy had wider implications for the state. With the economic and political weakness of the federal centre, and with the fate of regional leaders tied to regional elections, there seemed little prospect of establishing a 'united' national interest to rebuild what by the end of the 1990s, and according to many economic and social indicators, was an ailing Russian state. Huskey, for example, notes that, in the wake of the August 1998 financial crisis, some regions began to erect protectionist barriers normally associated with international economic relations, not intra-state trade (Huskey 2001: 115). Rather than seeing the federal centre as a threat to a particular ethnic identity, many regions began to see the federal centre as a bureaucratic obstacle to their own economic development (Borisov 1999). This, and not concerns over national identity, lie behind Putin's attempt to rationalise Russian federalism and reassert the power of the federal centre over the regions.

Decentralisation arguably represented as big a constraint on presidential power as the aforementioned weak pro-government support in the State Duma. As such, Russian federalism provides an extended, institutional rationale for the creation of the post-Yeltsin party of power, while the threat of separatism and state break-up provided a powerful discourse justifying centralisation. Together they constitute the logic behind the all-national United Russia, a point illustrated well by former chair of the Committee on Federal and Regional Policy in the State Duma (2002–7), Viktor I. Grishin:

> We believe that the maintenance of territorial integrity is a very important role of United Russia . . . when there are people who think the same way at all the levels of power and when all these people accept the party's discipline, we

believe that this structure makes the county stronger. Party discipline plays no less a role than laws and regulations. So, we believe that an all-national political party, represented in all the regions and at all levels of power, enhances the country's integration no less than state power itself.

(Grishin 2007)

In the post-Yeltsin period, the party of power was transformed from a tactic aimed solely at securing State Duma representation for the federal government in a decentralised political system, to a strategic organisation charged with a number of tasks, including the integration of a diverse elite across a large territory.

The creation of the all-national party

The creation of United Russia represented one in a sequence of events that constituted the initial process of centralisation. As noted in the previous chapter, attempts at centralising power began almost as soon as Putin came to presidential office in 2000 with changes to the modus operandi of the Federation Council and the creation of seven territorial districts covering the Federation, staffed by individuals loyal to the Kremlin. In turn, the law 'On Political Parties' was also a key moment in the centralisation process, paving the way for the all-national United Russia by placing restrictions on the scope and registration of regional parties. It is no surprise that on the same day that Putin signed this law, on 12 July 2001, Unity and Fatherland founded the steering body charged with merging the two movements (Ivanov 2008: 78). All-Russia, the second component of the FAR bloc, headed by St Petersburg mayor Yakovlev and Tatarstan president Shaimiev, were drawn into the merger process too and in December 2001 United Russia held its third and final Founding Congress.

United Russia appeared at the end of 2001 as the first serious attempt to create an all-national party of power organisation in the Russian Federation, one that synthesised the existing organisations of Unity, FAR and, in time, other smaller party organisations. The process of amalgamating other parties into this structure began almost as soon as Putin was elected president in March 2000. On 27 May 2000, Unity officially transformed itself from a political movement to a political party, incorporating Our Home is Russia, PRES and the People's Patriotic Party at the same time (Barakhova 2000) and, from December 2001, the party of power continued to incorporate smaller parties. In the autumn of 2003, United Russia began talks with the Industrialist Party (Promyshlennaya Partiya) on the possibility of a merger and in October 2006 the party held its last congress before approximately 68,000 members became United Russia members (Zakatnova 2006). This process continues to the present moment, with the Agrarian Party merging with United Russia in January 2009, and in this sense United Russia has been effective in counteracting potential opposition by incorporating them, in the way theorised in Chapter 1.

In terms of creating United Russia's regional organisation, there is some suggestion that FAR, as a relatively well-developed organisation that was itself a conglomeration of regional parties of power, gave United Russia a significant head start in its party-building efforts in the regions (Godzimirski 2001). There is also a suggestion that the strength of FAR represented an ongoing challenge to the federal executive branch and so figured as a decisive factor in the creation of United Russia to counter this threat (Hale 2006: 237). However, the actual strength of FAR prior to the merger is by no means clear. Despite the fact that FAR was formed several months before Unity in the run-up to the 1999 State Duma election, the organisation was typically shallow with little coherence. Makarenko (2000: 126), for example, notes that FAR had many of the characteristics of former parties of power, such as Russia's Choice, Our Home is Russia and PRES. These features included an amorphous structure, low motivation of middle-ranked staff and a weak network of voluntary activists.

Regional parties of power, as mentioned in the next part of this chapter, were rather insignificant actors in the Yeltsin period and so, by the time of the merger, FAR did not represent any significant organisational threat to either Unity or the federal executive branch. One of the heads of the FAR election campaign in 1999 and United Russia State Duma deputy since 2004, Vladimir Medinskii, makes the following observation on this subject: 'There was no physical organisation in terms of a party organisation and structure. It was an election bloc. I actually think that Fatherland All-Russia started being formed as a party structure only after the elections' (Medinskii 2007).

What is clear is that United Russia did inherit a number of economic and bureaucratic groups following the merger that, if not organisationally coherent, were nonetheless engaged in a bitter struggle for power in 1999. This experience contributed to the 'psychological' difficulties of merging Unity and FAR, compounded by the practical problem of resolving leadership positions and redundancy; problems inevitably encountered when two organisations become one. It has been suggested, for example, that there were personal issues between Fatherland leader, Yurii Luzhkov, and Unity leader, Sergei Shoigu, which made the initial face-to-face contact between them an obstacle. Medinskii (2007), someone with close relations with the former Luzhkov administration, paints an interesting picture of the tensions between these two leaders. According to Medinskii, their personal animosity towards each other was so great that they had trouble agreeing on the venue to discuss the merger.

In any case, the problem of potentially competing elite groups within the newly formed United Russia explains the choice of the *siloviki* representative, Boris Gryzlov, as the eventual party leader and further emphasises the early importance of the party's integrative role. Gryzlov helped steady United Russia during its difficult infancy, effectively replacing the party's first leader, Alekandr Bespalov, in November 2002 when the former became chair of the party's Higher Council. Bespalov, who was elected chair of the party General Council and Executive Committee in December 2001, was deemed surplus to requirements by the Kremlin, apparently losing out in a personal rivalry with presidential First Deputy Chief of Staff, Vladislav Surkov (Tregubova and Petrova 2002). However, the details surrounding Bespalov are far from clear. There was speculation that his shared history with Putin, possibly as far back as the KGB, but certainly to St Petersburg in the 1990s, became a source of discomfort for the president, as too Bespalov's ambitions and poor management skills (Sadchikov 2003). Bespalov's appointment as Federation Council Senator in June 2002, which according to him was against his wishes (Nikiforova 2002), marked the informal end of his leadership, although it was not until the party's II Congress in March 2003 that statutes were changed to shift power to the party Higher Council and its chair, Boris Gryzlov.[2]

Because Gryzlov held the position of Minister of Internal Affairs (2001–3), he was legally prevented from taking up the formal leadership of the party until November 2004 and the party's V Congress, when he was elected to the newly formed post of party chair. In terms of voter appeal, Gryzlov, who headed the St Petersburg regional branch of Unity in 1999, was an odd choice for a newly formed party looking for rapid electoral success. Although younger than former Unity leader, Aleksandr Gurov, Gryzlov (born 1950) still contrasts vividly with the energetic troika of leaders that led the party of power in 1999, but from the perspective of integration he was an ideal arbitrator between competing party factions and a trusted member of the St Petersburg *siloviki*.

As for the reasons behind the merger, many Unity and FAR deputies cite shared ideology as a driver. In some ways, this official line is supported by the fact that, in the period 2000–1, both party factions in the State Duma shared similar voting records (see Remington 2005: 42–3). In reality, though, it is difficult to speak of shared ideology as an explanation for the merger because of a lack of any coherent ideology within either FAR or Unity. Perhaps the only consistent 'idea' between Unity, FAR and the subsequent United Russia is the statist theme and therefore a latent conservative ideology. For example, the FAR party programme in the run-up to the 1999 State Duma election contained numerous references to *gosudarstvenniki*, defined as those people concerned with preserving the territorial integrity of the Russian state and ensuring effective rule of law (Otechestvo Vsya-Rossiya 1999). However, there were also key differences between Unity and FAR, notably in economic policy, with the former tending to espouse more social rhetoric coinciding with Luzhkov's own left-of-centre views (see Luzhkov 2005). In the final analysis, the 1999 State Duma election was a straight fight for power between two temporary electoral blocs that had little time to prepare, exhibiting few genuine links with society and relying heavily on administrative resources.

Moving on from this official version of the merger, it does seem reasonable to elevate the 'Putin factor' as an important, if not decisive, part of this process. There is little doubt that the merger was sanctioned by the president and that his growing popularity persuaded many elites in both Unity and FAR of the advantage of merging their respective organisations. As noted in Chapter 2, alongside the institutional rationale for United Russia there is also the strong influence of the intentions of power-holders in and around the federal executive branch, including Putin, and his desire to establish a particular kind of political system.

It seems likely that Putin's style of leadership and understanding of the potential of political parties influenced the decision to create United Russia as a tool to integrate the regions. It is clear, for example, that Putin, in comparison to Yeltsin (see Chapter 2), had a very different party experience. Yeltsin was someone familiar with party life, a long-term member of the CPSU, someone who had practical experience of parties from the inside, but someone who did not always see their practical value. Yeltsin's own style of leadership was by no means organised, and, as Colton observes, 'the President's team was deficient in teamwork' (Colton 2008: 331).

In contrast, Putin's biography shows a man looking at parties from the outside rather than from the perspective of a party politician. One of Putin's early professional encounters with political parties was as a KGB officer serving in East Germany, where he was responsible for collating data on a number of party organisations in NATO countries (Gevorkyan *et al.* 2000: 69–70). In the 1990s, Putin was active in the St Petersburg branch of Russia's Choice and then, in 1995, was responsible for managing the St Petersburg branch of Our Home is Russia, although, again, as someone operating largely behind the scenes, not publicly involved.

From the moment Putin assumed the post of prime minister in 1999 he was to have a much more direct relationship with the party of power. Unlike the parties of power of the Yeltsin period, which were parties of government ministers, such as Gaidar's Russia's Choice and Chernomyrdin's Our Home is Russia, first Unity and then United Russia became the party of Putin, so making explicit the link between party and president. This is not to say that Putin's personality or background predisposed him to any emotional attachment to either Unity or United Russia. Putin's leadership style, as both president and prime minster, is calculated, practical, business-like rather than intuitive. If Yeltsin tried to be a monarch, then Putin tried to be a manager, a 'CEO or Russia inc.' (Shevtsova 2005: 235), and, overall, Putin's attitude is summarised well by *Ekspert* journalist Aleksandr Mekhanik, representative of the liberal-leaning Club 4 November think tank:

> Putin is a system person; he wants to build a system with institutions, with pyramids, in order to manage society with less effort. He is also an educated person . . . he lived in Germany and saw how Germans were organised. So, he knows that democracy demands parties and believes that parties are a necessary feature of democracy, so Russia needs parties. But, where can we find them in Russia? There aren't any parties in Russia. Should we wait until they develop by themselves from grass roots? This may take hundreds of years.
>
> (Mekhanik 2007)

Importantly, Putin is also someone well aware of the problems in the regions, having served as head of the Presidential Monitoring Administration and then as Deputy Chief of Staff in the presidential administration with special reference to provincial affairs. Here he gained a first-hand insight into the weakness of existing structures linking Moscow with the regions (Huskey 2001: 118). Perhaps more ominously he also had a realistic view of lawlessness and corruption in many regional administrations and so the need to impose some kind of discipline. This point is relayed well by Yabloko leader, Sergei Mitrokhin, who recalls a meeting with Putin to discuss the case of Larisa Yudina, editor of the *Sovetskaya Kamykiya Segodnya* newspaper and member of the Kalmykiya regional branch of Yabloko, who was murdered on 7 June 1998 while investigating high-level corruption in the region:

> We had personal talks with Putin – Yavlinskii met him twice and I went to the meetings too. There were certain moments, for example, we talked about the situation in Kalmykiya. Larisa Yudina, the head of Yabloko regional organisation was killed in Kalmykiya. We asked for help in organising the investigation. But Putin said, 'What do you expect? Everybody is corrupt in Kalmykiya. What can I do? The system is completely corrupt'.
>
> (Mitrokhin 2007)

Putin's solution to the problem of weak central authority was to (re)create it using a number of parallel structures as layers of administration across the regions, including United Russia. The federal executive branch became the highest chain of command, but in psychological terms something akin to a 'leviathan', the ultimate state arbitrator keeping regional elites in check, creating the basis for political stability and a single, national interest.

United Russia as an integrating force

To understand United Russia as an integrative force, it is important to look at two related aspects of the party. The first is the ability of the party to attract candidates to run in elections and members to join and participate in the activities of the party between elections. As much as anything, this relates to the kind of selective and collective benefits the party is able to offer as well as their effectiveness as a means of control. This subject is discussed in more detail in the next part of this chapter. The second aspect of integration is the formal organisation of the party and its physical presence across the Russian Federation. This organisation – its branches, membership, leadership, links with other organisations and links with society – are all elements of the party's integrative capacity, and this is now discussed in the material that follows.

The formal organisation

The first regional branches of United Russia were established soon after its third Founding Congress on 1 December 2001. From this moment onwards, United Russia embarked on a rapid, party-building exercise that, within a six-month period, saw offices open in almost every region. The first regional branches were actually opened on 14 December 2001 in Moscow City and the Republic of

Adygeya, but by 22 May 2002 the party had branches in all but one of Russia's 88 regions (Edinaya Rossiya 2011b). Ongoing changes to Russia's federal structure between December 2005 and March 2008 reduced the number of regions from 88 to 83, but United Russia had already formed a physical presence in all of them by September 2003.

Along with a rapid party-building exercise, the new party of power also embarked on a rapid membership-building exercise, and, by the time of the December 2007 State Duma election, United Russia's membership surpassed the two million milestone, although since then there has been a change of emphasis from quantity to quality of members. In March 2009, the party introduced new regulations for prospective members, including a six-month probationary period and a more comprehensive application procedure (Edinaya Rossiya 2009d). Prospective members also need a prior recommendation from the Council of Party Supporters before their application can be considered. Party membership at the end of 2010 stands at 2,073,772 registered members.

According to official party sources (Edinaya Rossiya 2008a), the typical United Russia member tends to be female (60.3 per cent), white collar (59 per cent), in the 40–65 age range (49.6 per cent), with a higher education (39 per cent) and working predominantly in the sectors of education (28.1 per cent) and industry (30.5 per cent). As discussed in Chapter 3, United Russia's membership is largely symbolic in the sense that it does not form a major plank of party financing – in contrast to genuine mass-parties (Duverger 1964: 63). However, membership bestows a great deal of legitimacy on the party, which, in practical terms, enhances the party's electoral appeal, as the following quote explains:

> Our party goal has three aspects: [first] attract voters through a large number of [party] 'good deeds'; [second] attract party members, which is also important because a large membership shows that the party is attractive to people and we work intensively on this and [third], of course, propaganda and the dissemination of information on the party organisation and its work.
>
> (Svyatenko 2007)

In terms of organisational structure, United Russia does not differ greatly from any other Russian party, relying on what may be termed a 'centralised party model', corresponding with the general experience of party-building in the post-Soviet period (see March 2002; D. White 2006). This model is characterised by the subordination of regional branches and rank and file members to party central office, in the Russian case the Central Executive Committee and the federal-level party leadership. For United Russia, this pyramid-structured organisation is superimposed over the existing territorial boundaries of the Russian Federation, identical in form from one region to another, with efforts made from the outset to ensure this conformity.

According to the party statutes (Edinaya Rossiya 2009e, article 13.3) a maximum of one regional branch is allowed in any one subject of the Russian Federation, giving local branches a catchment area that follows existing municipal boundaries. Primary party organisations (*Pervichnye*), as the smallest subdivisions, are governed by party municipal and regional offices. What is interesting about this structure, at least from the perspective of integration, is that the party's Interregional Coordination Council was originally organised along the lines of the seven military districts introduced as administrative divisions by Putin at the very beginning of his first term as president in 2000, and which represented the first moves in the centralisation process. Following Dmitrii Medvedev's decree (21 January 2010) to create an additional North Caucus district, United Russia's Presidium of the General Council quickly followed suit creating its own North Caucus Coordination Council on 17 February 2010 (Edinaya Rossiya 2010f). The Central, North-Western, Southern, Volga, Ural, Siberian, North Caucus and Far Eastern districts therefore represent an agglomeration of regional party offices and the next level in the vertical chain of command before the federal-level executive bodies of the party.

The breakdown of United Russia's ruling bodies, according to the latest party statues (21 November 2009), is as follows. At the apex of the party is the position of party chair, which from April 2008 has been occupied by Vladimir Putin. Boris Gryzlov, the former party chair, regained his previous position as chair of the Higher Council following Putin's arrival. The next leadership positions in terms of their authority are the two co-Chairs of the Higher Council, currently occupied by former Unity leader Sergei Shoigu and former All-Russia leader Mintimer Shaimiev. Shaimiev retains his position in the leadership of United Russia despite stepping down as president of Tatarstan in 2010. Former Fatherland leader, Yurii Luzhkov, another original founder of United Russia, was the third party co-chair until October 2010. However, his high-profile clash with President Medvedev resulted in his dismissal from United Russia. Party leaders have a four-year mandate, appointed through open election by the party congress.

The Bureau of the Higher Council represents the next ruling body, which includes chair and co-chairs of the party Higher Council, the secretary and co-chairs of the General Council Presidium and the head of the Central Executive Committee and 'others' (article 7.3.5.3). This bureau is largely responsible for personnel decisions and for making recommendations to the regions and the congress, but also for approving party-lists. The party Higher Council is the next ruling body in the party hierarchy, comprising 68 individuals and responsible for deciding the general direction of party development. The party Higher Council, according to party statutes, is elected from the most eminent public and political figures in the Russian Federation, enjoying authority both within Russia and also on the international stage.

Subordinate to the Higher Council is the Presidium of the General Council, followed by the General Council, which together have the power to appoint the Interregional Coordination Council and also the Central Executive Committee. The Interregional Coordination Council operates as an extra layer of administration within the party, overseeing the activities of party branches within Russia's seven territorial districts, acting as an interlocutor between the federal and regional party organisations (article 13.12). The Central Executive Committee is responsible for developing party ideology, in line with the decisions of the party congress and party programmes, in line with the decisions of the party congress and Control-Revision Commission has responsibility for the implementation of decisions by the ruling bodies and also bodies. The Central activities of the party executive (article 12.1). Figure 5.1 statutes, the implementation of these federal-level ruling bodies. al and economic an organigram

Although regional party offices have a degree of autonomy to conduct their election campaigns (see Chapter 3), the party is, at least in forms, very centralised with a clear hierarchical command and control structure, emanating downwards from the party chair. In principle, all party factions in regional parliaments are subordinated to their respective regional par's 83 The party faction in the State Duma, as the federal-level parliament, is subject to the Central Executive Committee. Regional parliaments are able to communicate with their colleagues in the State Duma and even propose federal legislative initiatives, which if taken up by the United Russia faction have the chance to become federal law. However, and as the previous chapter noted, it is unlikely that any regional legislative initiative will be adopted by the party faction in

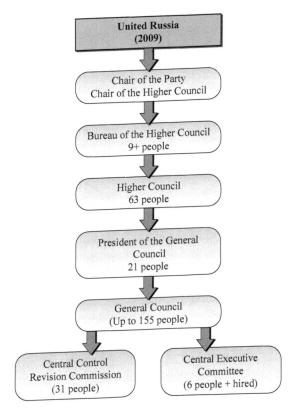

Figure 5.1 United Russia federal-level structure. Source: Edinaya Rossiya (2009e).

agreement of government and the federal executive
the State Duma United Russia is part of the centralisation process that
branch. As alre... ver of the regions vis-à-vis the federal centre and to avoid
aims to decrea for certain regions, as was the case with the asymmetrical
preferential eltsin period. For this reason, it is no surprise that the party
federalism nity, with one set of party statutes governing the entire party
strives fo
organisⱭ this formal organisational set-up, the degree of centralisation within
CoⱭsia is difficult to gauge. In many respects, this is due to the mismatch
Unit formal and informal procedures that are a feature of any organisation,
be ticularly so in Russia. Despite the obvious prominence of the federal-level
b rship, the party congress is the highest ruling body of the party, confirmed
the party statutes, but also by article 24 of the law 'On Political Parties' (CEC
2001), which also lays down stipulations governing party life, stating that 'the
supreme leading body of a political party shall be the congress of the political
party'.

United Russia's congresses are scheduled by the Presidium of the General
Council and, according to the party statues, must take place at least once every
two years, but only with the approval of the party chair and other federal-level
ruling bodies (article 8.3.1). The congress is formed from regional party representatives and has numerous jurisdictions, including the ability to change the
structure of the party, undo previous party decisions, change the party statutes,
approve party programmes and elect party leaders, including the party chair, chair
of the Higher Council, members of the Higher Council, members of the General
Council, members of the Presidium of the General Council, the Central Control-
Revision Committee and the party nominee for elections for the president of the
Russian Federation (article 8).

In practice, however, there is reason to doubt the independence of the party
congress from the federal-level leadership. As United Russia's regional-level
organisation should, in theory, mirror that of the federal-level organisation,
regional party congresses or conferences should have the final say in resolving regional leadership issues. However, there is a suggestion that the party's
federal-level leadership has a strong influence over appointments in the regions,
recommending candidates before the congress votes, turning intra-party elections
into a closed contest. The following quote by a United Russia deputy recalls his
experience of a regional leadership election: 'People from the federal General
Council came to oversee our elections too. They gave us their recommendations and said, 'let us elect this or that person'. Then we voted for the selected
candidates' (Tyagunov 2007).

In comparative terms, a typical indication of the actual degree of centralisation
within a party can be seen in the process of drawing up candidate lists for elections. If party-lists are determined locally, then this suggests decentralisation of
power within the party, but if these lists are subject to approval and amendment by
e national leadership, then this supports the existence of centralised party power
zerbiak 1999: 533).

Drawing up United Russia's party-lists is a multi-stage, meticulous process. Again, in formal terms, the composition of party-lists for federal-level elections should be decided by regional primaries, where lists of candidates are finalised according to the popularity of the candidates in question, determined through consultation with party rank and file and the wider public. The greater the candidate's popularity rating in these primaries, the higher their eventual place on the party-list. The total number of places on the list reflects the circumstances of the individual regions, such as the overall number of registered voters, but also the number of party members that the regional branch has succeeded in attracting.

The 'hidden hand' of the federal-level ruling bodies can be seen in the final outcome of these party-lists. It is widely reported that party-lists are sent to Moscow so that candidates can be scrutinised and vetted, especially for links with criminal organisations. Although the regional congress should decide these issues more or less autonomously, in many cases (but not all cases; see the final section of this chapter) the federal-level party leadership exerts considerable influence. As United Russia deputy, Aleksandr Tyagunov, comments, the party leadership often recommends that regional branches implement certain procedures, but it is unclear if recommend is a Soviet-style synonym for 'command': 'Today, we don't force people to do things, today we recommend they do . . . some recommendations are very tough, some are soft, but yes, the party monitors if they have been fulfilled' (Tyagunov 2007).

Socialisation

Beyond United Russia's formal organisation and membership, the party is also active in supporting wider societal initiatives and directing resources to cultivate and support them to raise the profile of the party among the electorate. As already noted, the party has enjoyed some success in incorporating smaller parties and so acquiring extra linkage with society as a whole. By December 2008, United Russia had established formal cooperation agreements with over 70 public associations of varying sizes, each with their own grass-roots networks (Edinaya Rossiya 2008b). These comprise a fairly disparate group of sports associations and business associations and a significant number of national-cultural autonomies, reflecting Russia's ethno-national diversity and including the Association of Chechens, the Union of Georgians in Russia, the Association of Ukrainians in Russia, Tartars of Russia and several organisations representing Cossacks, among others. United Russia also has agreements with the Oil, Gas and Construction Workers Union and in April 2008 signed an extensive cooperation agreement with the Federation of Independent Trade Unions of Russia (Shkel' 2008). This union signed a similar agreement with United Russia in 2004 and by October 2010 included 48 all-Russian trade unions and 79 regional trade unions (FNPR 2010). The union continues to have a direct communication channel to the party through the Duma chair of the Labour and Social Policy Committee, Andrei Isaev.

The Orthodox Church is another strategic partner for United Russia – the party keen to exploit the mobilisation capacity and legitimacy of this older, more respected institution. In broad terms the Church subscribes to the stability emphasised by the post-Yeltsin regime, making the Church–United Russia nexus quite logical. United Russia's exact relationship with the Church appears to be at a number of levels. There are, undoubtedly, supporters of the Orthodox Church within United Russia (and A Just Russia) who may be considered practising Orthodox Christians, but the party also has contacts with the Church through informal meetings between clergy and party leaders and has used this connection to generate electoral support and to publicise some of the party's 'real achievements'. One of United Russia's ongoing projects, *Istoricheskaya Pamyat'* (historical memory), aims to restore buildings of historical significance to their former, often pre-Soviet, glory, including a number of Orthodox Churches (Edinaya Rossiya 2010g). Again, the exact input of United Russia in this project is unclear, although it appears that the party coordinates funds from the state and private business to help restore these buildings – fund-raising on behalf of the Church.

In fact, many of the organisations that have strong links with United Russia also appear to be components of the power vertical and so additional layers of administration across the regions, like the party. In this sense, United Russia is only one structure used to integrate. This echoes the conclusion from Chapter 3 that United Russia, although the most important party agent of the federal executive branch, is by no means the only one. In this sense, integration through a vertical chain of command is complex, exhibiting a horizontal dimension and containing numerous elements including, but by no means limited to, political parties.

The prominent United Russia think tank, the Centre for Social Conservative Policy, is one of these all-national structures. In fact, this organisation was not created by the party, or even expressly for the party, as a think tank, but as an organisation to monitor public opinion and public reaction to certain laws. Roman Skoryi, director of international affairs for the organisation, recalls the reasons for its creation:

> The Federal Law 122 [monetarisation law, see Chapter 6] was the trigger for establishing our centre. After the law was adopted and implemented people went to the streets, so it became clear that it was necessary to create an analytical centre that would be able to investigate the situation inside the country and to assess whether some draft laws would be possible to adopt and implement under certain situations, or if it is better to wait until the situation in the country changes.
>
> (Skoryi 2007)

This organisation, like many others associated with the party of power, has an impressive all-national organisation of its own and provides an important forum for debate and for party training. Although a great deal of party training is focused on electioneering, the ability of United Russia to bring elites together from across the regions to participate in party events contributes towards the sense of common purpose that the party is undoubtedly trying to create. In general, training the party cohort is taken seriously within United Russia and, as mentioned in Chapter

3, there are many indications that this party of power, more than any other, can be regarded as a genuine 'learning organisation' (Holt *et al.* 2000), or one able to adapt and succeed in its environment.

Many party workers, especially younger employees, are given experience in public organisations before they start work in United Russia in order to improve their effectiveness, but also to develop the party's links with these organisations. Institutions such as the Moscow School of Political Research and the Academy of the State Service, among others, provide training and expertise in areas of election campaigning and the use of PR technologies. United Russia also organises pre-election training for its activists, inviting professionals, heads of PR agencies and other well-known PR specialists to conduct training sessions. Party members receive hands-on training to equip them with the essential electioneering skills needed by the party, such as how to interact with television cameras, give public speeches and conduct press conferences, all important for the party's media-reliant strategy, which is crucial for the party's electoral success.

However, this training also has a direct relationship with integration. A prominent journalist writing for a national newspaper and someone familiar with United Russia and its party-building efforts, makes the following observations on the effectiveness of United Russia as a medium for socialising elites from across the regions:

> They invite party functionaries from the regions to the party central office, where psychologists, political technologists, journalists and politicians work with them. They tell them how they should behave, how they should act, what is good and what is bad, what the party expects from them and things like that. In this way they develop a kind of unified standard of behaviour and views. This training resembles dog training – maybe it sounds a bit rude, but it looks like that. Nevertheless, this training over the years has done its job and the party is getting more and more unified across the country.
>
> (Anonymous 2007)

United Russia is also keen to learn from history, including comparative party experience, in order to unify Russia's political space. Material from Gleb Pavlovskii's Europa publishing house together with in-house party publications provide activists with an idea of the party's larger mission. The in-house series *Biblioteka Edinoi Rossii* (United Russia Library), for example, includes overviews of party-building experience in a number of democracies, including the British Conservative Party, French Gaullists, Japanese Liberal Democrats and German Christian Democratic Union and Christian Social Union (Frolov *et al.* 2003). These parties provide useful points of reference for United Russia, as parties that oversaw periods of reform in their respective countries but that also enjoyed periods of sustained electoral dominance. On 1 December 2009, the party launched its 'world experience' (*mirovoi opyt*) project aimed at studying the efforts of other parties in overseeing successful conservative modernisation (Edinaya Rossiya 2009f).

In recent years, the party has also attempted to establish contacts with a number of international partners. By November 2009, the party had signed agreements of cooperation with 20 foreign parties, mostly 'ruling' parties in Russia's near aboard, but including smaller parties in countries with a Russian-speaking minority (Baltic states) as well as parties in the de facto republics of Abkhazia (Edinaya Abkhazia – United Abkhazia), South Ossetia (Edinstvo – Unity) and Transnistria (Obnovlenie – Renewal). The party also has good contacts with the Japanese LDP and the Israeli Kadimi (United Russia even opened a mission in Israel in August 2008) and in March 2008 Konstantin Kosachev (the deputy chair of the Presidium of the General Council responsible for contacts with foreign political parties) announced that the party had successfully joined the International Conference of Asian Political Parties (Edinaya Rossiya 2008c). As of 2010, Kosachev represents United Russia in the organisation's standing committee.

United Russia has also been keen to incorporate the substantial party-building experience from the Soviet period. The Central Executive Committee, which hires professional staff to work on party projects, includes several former members of the Executive Committee of the CPSU. Their experience of Soviet *apparat* culture has proved a valuable source of know-how for controlling and monitoring the fulfilment of party decrees and goal attainment across such a large territory, along with assisting in other aspects of party management (Il'nitskii 2007).

In the post-Yeltsin period, there has also been a discernible proliferation of youth organisations, enabling United Russia to extend its socialising capacity to these organisations too. One such organisation, United Russia's semi-official youth organisation, Molodaya Gvardiya (see previous chapter), was formed in November 2005 as a replacement for the outdated Molodezhnoe Edinstvo. As a youth organisation oriented towards social tasks, such as collecting litter and painting benches, etc., Molodezhnoe Edinstvo was considered out of touch with both Russian youth and the increasingly political demands of the new regime. In contrast, Molodaya Gvardiya is an explicitly political youth organisation that aims to advance young people into mainstream politics. The phenomena of *PolitZavody* (political factories) were created to achieve this goal and to exploit the decision taken by United Russia in April 2006 to introduce a 20 per cent quota for under 28 year olds on the party's electoral lists. *PolitZavody* are effectively primaries for young people conducted in every region of the Russian Federation consisting of three tours that enable Molodaya Gvardiya to select suitable candidates who could eventually make United Russia's party-lists. However, there is no requirement that an individual belongs to Molodaya Gvardiya in order to participate in this process (Molodaya Gvardia 2009b). Members of other youth organisations, including radical and opposition organisations, are also free to participate, indicating that this is very much a strategy designed to attract as many young people as possible.

Molodaya Gvardiya, like United Russia, is also all-national in scope and so provides a useful medium for the party to attract and integrate young people from across the regions and to instil party values early on. The 'young guards' are also encouraged to attend courses and events with a distinctive ideological tone and to learn those important skills required for a future career in politics – namely how to effectively interact with the media. Accordingly, the dissemination of political material is considered an important part of Molodaya Gvardiya work, as too is supporting the general line of United Russia. Molodaya Gvardiya lists supporting the ideas of United Russia as one of its statutory purposes (article 1.3, Molodaya Gvardiya 2009a) and the value of this organisation for integration and, more generally, for the persistence of dominant-power politics is summarised well by Molodaya Gvardiya activist, Nadezhda Orlova: 'The initiative to create Molodaya Gvardiya came from the very top, this is a pro-power youth organisation, which means pro-Kremlin and pro-Putin and we openly declare that we stand for succession of the current political course' (Orlova 2007).

It is also important to note, as elsewhere, the web on pro-power interconnectedness. Ivan Demidov (born 1963), who assumed the leadership of Molodaya Gvardiya 2005–8, also has strong connections with both the media and the Orthodox Church, working on several television channels in the 1990s and playing a prominent role in establishing the Orthodox Christian television channel, Spas, in 2005 (Toropov 2005). Demidov briefly worked on ideological issues in United Russia (2008–9) and in April 2010 was appointed to work in the domestic politics directorate of the presidential administration, overseeing public and religious organisations. This move was seen by some as an indication of the increasing prominence of the Orthodox Church in Russian politics (Bolotova 2010).

Elsewhere, organisations with close connections to United Russia provide training for party members and representatives of pro-Kremlin youth organisations. For example, the Club 4 November think tank developed a 'school for young leaders' – a short-term training course designed specifically for young business people and lawyers with the aim of raising and disseminating ideology through workshops, role playing and other activities. According to a representative of the Club 4 November, research conducted by the think tank revealed that most Russians are what can be termed 'ideologically lost', so this kind of training supplies young people with the ideological concepts they are lacking, in particular those that make sense of the often contradictory concepts of economic liberalism within a socially oriented state. Aleksandr Mekhanik summarises the aim of these courses: 'First of all, the purpose is to spread ideology, to instil it in young people's consciousness so that ideology could take root there, in people's minds – as Vladimir Ilyich Lenin recommended. We still use his methods' (Mekhanik 2007).

At a very general level, United Russia's development of its personnel goes some way to promoting an overall political culture that, although rudimentary, nonetheless represents a first real attempt in the post-Soviet period to revise or replace the old Soviet political culture. Andrei Chadaev, former member of the Public Chamber and, as of 2010, deputy head of United Russia's Central Executive Committee, sums up this political culture in the following statement:

> United Russia is kind of a buffer in the fight between the centre and the regions built against the background of a silent agreement between the Kremlin and the regions, that is, 'Yes, we (the regions) fight for power and influence, but we do not use separatism in our fight'. Those who use separatism are outlaws.
>
> (Chadaev 2007)

United Russia is part of a new approach to post-Soviet Russian politics, one that grants elites the freedom to go about their political business, as in the Yeltsin period, but with certain limits imposed on their activities. These limits support the vision of Russian politics held by the statist element that currently holds sway in the post-Yeltsin period.

The centripetal pull of the party

The socialisation capacity of the party is just one aspect of integration, but this has relevance only if the party is successful in attracting supporters. This section looks in more detail at the kind of benefits the party is able to offer candidates and rank and file members and how they translate into an overall integrative potential. They include the collective or solidarity benefits associated with the party label and the selective benefits or private goods that United Russia affiliation provides. Collective benefits attract like-minded individuals who wish to maintain the party's label or ideology; selective benefits attract supporters who perceive material or career advantages from the party.

Collective benefits

United Russia, as a party label or brand, is extremely recognisable across the Russian Federation, due largely to the effect of favourable media coverage noted in Chapter 3. Voters are easily able to identify United Russia as a political party, but, more importantly, as the party of Putin, of Medvedev and of power in general. Hale and colleagues suggest that, even by the 2003 State Duma election, a majority of Russians perceived a strong link between Putin and United Russia, with 80 per cent of respondents expressing an opinion that Putin fully supported United Russia and 84 per cent that United Russia fully supported Putin (Hale *et al.* 2004: 294). By 2009, an open-question survey ($n = 2,000$) conducted by the Public Opinion Foundation (FOM 2009) inviting respondents to name United Russia's leaders showed that 70 per cent of respondents successfully identified Putin, but that, interestingly, Dmitrii Medvedev was the second most mentioned party leader, with 29 per cent of respondents mistakenly naming the president as a party leader. Only 19 per cent of respondents named the long-serving Boris Gryzlov.

Despite voters making the (intended) connection between party and power-holders in the federal executive branch, many struggled to identify the party's ideology and policy positions and this is of consequence for understanding the strength of United Russia's solidarity benefits. A survey conducted in 2006 in over 40 regions showed that 44 per cent of respondents had a positive attitude towards the party, but only 5 per cent had a positive attitude because they 'understood the party's goals and liked its programme and policy' (Klimova 2006). Five similar surveys conducted between June 2006 and April 2008 showed that, of those respondents expressing a positive attitude towards the party (an average of just over 50 per cent of respondents), only 4 per cent cited the party's programme as the reason for this positive attitude in a follow-up open question

(FOM 2008b). At the same time, an open-ended survey from 2008 saw only 3 per cent of respondents cite programme and ideology as a reason for United Russia's popularity (FOM 2008a).

As discussed in Chapter 3, United Russia has a catch-all electoral strategy, one that retains a fuzzy ideological focus in order to appeal to as many voters as possible. Party programmes and manifestos tend to contain a great deal of rhetoric or valence issues, unlikely to alienate voters. As a result, identifying United Russia's collective or solidarity benefits is a much more complicated task than for the corresponding particularistic or selective benefits.

In general terms, it is possible to discern some ideological or quasi-ideological threads that give some indication of the strength of the party's collective benefits. As noted in Chapter 3, the ideological orientation of the party rests on three 'isms' – centrism, Putinism and conservatism – that have been consistent themes in the politics of the post-Yeltsin period. The first thread, centrism, is more a commitment to avoid ideological extremes and is, in many ways, indistinguishable from the aforementioned electoral strategy of catchallism. As such, centrism has the power to attract candidates who themselves have a fuzzy ideology as well as those career-minded politicians who believe in subordinating ideology to the expediencies of power.

In view of the aforementioned party origins and reversed principal–agent relationship, United Russia's party label is closely associated with a second ideological strand, termed here 'Putinism'. As discussed previously, the Putin factor continues to figure strongly in United Russia's development, and even in ideological terms the influence of Putin and the federal executive branch remains strong. Vladislav Surkov, an influential figure behind the scenes of United Russia, has been instrumental in developing the broad contours of the post-Yeltsin regime's ideology, as well as the ideas associated with Putinism. He was appointed Deputy Chief of Staff in August 1999 when he began to work on domestic politics, including United Russia. One of Surkov's ideological tasks has been to lessen the party's reliance on the personality of Putin by creating a more durable 'ideology of Putin' and, again, this diversification strategy, of reducing the party's dependence on a single personality, mirrors the strategy of creating multiple party alternatives to United Russia. Surkov is on record as saying that the greatest flaw in the Russian political system is its reliance on the resources of one person (Putin) and, consequently, its reliance on one party (Surkov 2006a).

As a result, Putinism as a collective party benefit operates at two levels, each with a slightly different emphasis, but each able to attract supporters. At the first level, Putinism is an embodiment of those qualities associated with a strong leader and authority figure, the party essentially becoming an extension of Putin's own personality. This level of Putinism has no intellectual substance and is merely a continuation of the personalised politics that accompanied the establishment of presidentialism in Russia and the collapse of the Soviet Union in general. A twenty-first-century 'cult of personality' is perhaps an extreme characterisation of this level of Putinism, one that misses the important point that the Russian public (perhaps mistakenly) tends to perceive state officials such as the president as elected figures who come and go (Kostikov 2007) rather than as 'lifetime leaders'. The second level of Putinism represents a more recent attempt to develop this personality politics into a coherent set of ideas, which as a broad world view may have some ideological claim. One important idea that emerged during Putin's second term of office was the notion of 'sovereign democracy', used to describe the particular (non-Western) development of democracy in Russia in the post-Yeltsin period.

The tenets of sovereign democracy have received a number of interpretations in both Russian and English language literature, reflecting the fact that neither the Putin nor Medvedev administrations have provided much detail on this subject. These tenets include the basis of sovereign democracy in traditional Russian values: the need for Russia to follow its own style of democracy, to maintain economic growth and to preserve Russia's borders or sovereignty (Tret'yakov 2007). In short, sovereign democracy was an attempt to (re)-emphasise the undoubted variation that exists among existing democratic states and to move away from the notion of a one-size-fits-all democracy.

In comparative terms, there are parallels between sovereign democracy and styles of communitarian democracy, or what Katz terms 'collectivist popular sovereignty theories revisited' (Katz 1997: 79). In particular, sovereign democracy resembles 'guided democracy' seen in Indonesia under President Sukarno, as well as 'basic democracy' in Pakistan under Ayub Khan and 'national democracy' in the Republic of Guinea under Sekou Toure. In the case of Indonesia, guided democracy (1959–65), like sovereign democracy, also played on the importance of unity and promoted order (primarily stability) and national development (primarily economic growth) through consensual, as opposed to conflictive, democracy (Brooker 1995: 181–90; Katz 1997: 89). Guided democracy also legitimated a limited role for opposition vis-à-vis the ruling party, Sekber Golkar, which served as the civilian front for the military-controlled regime.

In many ways, sovereign democracy justifies dominant-power politics by providing a broader intellectual basis to legitimate the actions of the ruling elite. Sovereign democracy must also be placed in the context of regime change and colour revolutions occurring in the period 2000–4, and the need to re-emphasise a democratic discourse to reduce the discursive space available to regime opponents. This is not to say that sovereign democracy has been discarded after fulfilling this purpose. Although Medvedev once labelled sovereign democracy a 'far from ideal term' when he served as deputy prime minister (Medvedev 2006), figures within the presidential administration, notably Surkov, continue to use the term. However, the reality is that nearly all of the intellectual output from the federal executive branch simply builds on one overarching process that has thus far underscored the post-Yeltsin period, as government representative in the State Duma, Andrei Loginov, comments: 'Building the power vertical and building United Russia were two aspects of one single process. This was the process of creating stability and enhancing state control over social processes in the country' (Loginov 2007).

In this respect, Putinism is primarily about stability and this appears to strike a great deal of resonance with ordinary voters and, so too, many United Russia members. In a detailed survey conducted by VTsIOM (2009) on the Russian national idea, respondents were asked to rate Russian and Soviet leaders from the past 100 years on their success or failure in leading the state in the right direction. Yeltsin (64 per cent) and Gorbachev (63 per cent) received the top two negative responses, while Putin (80 per cent) and Brezhnev (41 per cent) led the positive estimations. Research from 2008 cited in the same VTsIOM report also revealed that 55 per cent of respondents supported stability and reform of an 'evolutionary character'. In terms of the bigger party picture, United Russia dovetails well with this desire for stability, as an attempt to counteract the centrifugal tendencies in the regions and to reaffirm the authority and integrity of the state. As such, stability reflects a third underlying ideological strain within United Russia, but clearly the ideological basis of dominant-power politics – 'conservatism'.

In fact, the conservatism of the post-Yeltsin period is by no means a new or separate development, but is another idea that predated the arrival of Putin by some way. Although figures such as Surkov have worked hard to supply the regime with ideological content, many of the ideas in vogue in Russia since 2000 were already fairly mature by the end of the Yeltsin period. For example, Vyacheslav Nikonov and Sergei Shakhrai, both key figures in PRES, published a Conservative Manifesto in October 1994 (Nikonov and Shakhrai 1994) that stressed similar themes of preserving traditions, building democracy through strong institutions, a commitment to economic development and private property and warnings of the danger of asymmetrical federalism.

As such, the ideas associated with conservatism, including stability, are probably the strongest collective benefit that the party offers. It is worth noting that, by the end of 2010, the biggest party competitor to United Russia remains the CPRF and, in the post-Yeltsin period, the median Russian voter still gravitates to the left of centre, so United Russia's emphasis on social conservatism reflects this general tendency in Russian society. It is partly for this reason that Boris Gryzlov declared Russian conservatism as the party's official ideology at the XI Congress in St Petersburg in November 2009 (Edinaya Rossiya 2009g), with the 'Russian' element representing socially oriented 'care of the Russian people'. This development represented a victory for the dominant wing within the party and its long-standing attempt to legitimate the party through a traditional ideology, such as conservatism. Since 2004, this dominant wing has tried to establish durable ties with the European People's Party (EPP) – a prominent group of centre-right parties spread across EU and non-EU states – but so far these attempts have failed. At the time of the party XI Congress in November 2009 an NTV news broadcast claimed that the party had indeed joined this prestigious party family (Stoev 2010), although this report was later found to be erroneous.

United Russia's conservative ideology, and broadly that of the post-Yeltsin period to date, represents the most discernible ideological content of the party label and so the most effective means of attracting and integrating supporters.

The following comments by a United Russia deputy in the Moscow Legislative Assembly summarise this point clearly: 'I believe that United Russia, being by nature a centrist party, is strongly attracted to conservative ideas that combine, on the one hand, ideas of strong state and, on the other the development of private initiative' (Krutov 2007).

However, this brand of conservatism is not without its problems. It is clear that the social orientation of Russian conservatism with its adherence to social protection does not fit well with many of the market-oriented laws passed by the party faction in the State Duma, including the hugely unpopular monetarisation law. There is also a question mark against the congruence of conservatism with the progressive, even liberal, economic development envisaged in Medvedev's modernisation agenda or even the revived Strategy 2020. It is no surprise that, immediately following the declaration of conservatism at the November 2009 congress, reports of a rift between party and president appeared in the press (Bilevskaya 2009).

At the same time, the strength of the party's collective benefits is undermined by the fact that most United Russia parliamentary factions exhibit a great deal of ideological diversity that makes it difficult to definitively characterise the party as conservative. Within the State Duma or any of Russia's 83 regional legislatures, the United Russia faction invariably contains a diverse mix of individuals who hold a range of views and who publicly or privately, previously or currently, are monarchists, Orthodox Christians, conservatives, laissez-faire liberals, democrats, socialists and even anarchists, among others. Although most United Russia deputies claim that they join the party to make some contribution to improving the lives of ordinary Russians, it would appear that United Russia offers stronger selective benefits as the party closest to power. A comment by a United Russia deputy underlines this point:

> I don't want to offend anybody, but it's clear that in reality, the process of forming political parties has only just started in this country . . . The party is very heterogeneous. I think that many people – I want to be honest with you – joined the party just as a career move.
>
> (Anonymous 2007)

In the final analysis, United Russia is probably the first attempt to unite the (often contradictory) ideas of liberal reform, social justice, national pride and patriotism (Fedorov 2007: 198), but beyond support for stability and Putin the United Russia label offers few clear footholds for supporters. This lack of ideology is captured well by the foreign partners the party has so far courted. A senior journalist recalls his surprise at the list of (19) foreign parties invited to the party's VII Congress in Ekaterinburg in December 2006:

> Their Round table seemed to be the only place in the world where representatives of Republican Party of the USA were at the same table with representatives of the Communist Party of China. At that time, the Republicans were

the party of power in the USA. If it had been the Democrats, they would have invited the Democrats!

(Anonymous 2007)

Selective benefits

United Russia as a large and well-funded organisation is able to offer a number of benefits to candidates, including financial resources necessary to pay for advertising, campaign staff, pollsters, transportation, etc. In addition, United Russia also offers expertise, training and other organisational benefits mentioned previously in this chapter. In short, any candidate interested in a successful career will no doubt view United Russia as currently the best option available and the absence of any strong ideology simply adds to its attractiveness for many career-minded candidates.

The ability of the party to utilise its financial advantage to attract supporters corresponds neatly with the favourable economic conditions that have accompanied the rise of dominant-power politics in the post-Yeltsin period. As already noted, control over financial flows is a factor in the strength of the regime, but also in the party of power too, as the preferred party agent of the regime. The increase in state revenue from oil and gas sales has directly enhanced the party of power's ability to attract candidates and integrate them into the party organisation.

However, an important and related development in terms of the party of power's selective benefits is the ongoing process of centralisation, which has removed many of the alternatives available to candidates wishing to compete in elections. As noted in the previous chapters, before 2007, half of all State Duma seats were decided on the SMD portion of the vote, meaning that many candidates ran independently without any party affiliation. In many respects, the switch to the 100 per cent party-list election format was another step along the path of centralisation, forcing candidates to commit to controlled structures, such as United Russia, to continue their political careers. This increase in the importance of party resources, notably party affiliation, did not entirely coincide with the abolition of the SMD portion of the vote in 2007 (Golosov 2004; Hale 2006; Smyth 2006), but has been a tendency evident throughout the post-Yeltsin period, as the control of the federal executive branch over political processes has grown.

Evidence for the link between centralisation and increasing party selective benefits can be seen in the way that regional parties of power were rare occurrences in the decentralised conditions of the Yeltsin period, pointing to the fact that parties offered few resources to incumbents in pursuit of their career ambitions, either political or economic in nature. Golosov (2004) devotes a whole chapter to the formation of political parties in Sverdlovsk Oblast, including the prominent regional party of power, Transformation of the Urals, established in November 1993. However, Sverdlovsk was, in many ways, an exceptional case, as in most regions parties were largely seen as irrelevant by regional executives who had access to all the resources they needed by virtue of their institutional affiliation – party affiliation carried no extra advantage. In the case of Sverdlovsk, the incumbent governor, Eduard Rossel, was dismissed from office by Yeltsin in November 1993 and so challenged for executive office from a legislative position, making a party vehicle more expedient in this case than in many others (Golosov 2004: 134). Although this regional movement was a useful means of mobilisation, Rossel's personal political resources were still more consequential (Gel'man and Golosov 1998: 35) and, according to the general definition of the party of power offered in Chapter 2, Transformation of the Urals was not a party of power in the sense that it was created as a platform to challenge executive power rather than support it.

In the post-Yeltsin period, the strength of the federal executive branch enabled centralisation and centralisation reduced the space available for political entrepreneurs. At the same time, the strength and so the attractiveness of United Russia continued to grow. This is not to say that previous parties of power had no success in attracting high-level regional candidates to their cause in the decentralised conditions of the Yeltsin period. High-status regional elites often ran on their party ticket or supported them in other ways. During the 1999 State Duma election, even when there was a great deal of confusion over the identity of the next president, the former party of power, Our Home is Russia, still managed to muster an impressive array of local notables, including six governors as well as a number of other high-ranking regional officials (Panorama 1999). As detailed further in this chapter, supporting the party of power was, and still is, a way of demonstrating loyalty to the president and of enhancing individual and regional bargaining positions with the federal centre. What is different in the post-Yeltsin period is that United Russia is attractive to many candidates because of the coattail effect from association with the popular president and, from 2008, the popular prime minister.

As a result, United Russia offers ambitious individuals the selective benefit of the party label, although this benefit is arguably a collective benefit in the sense that it is universally available to all candidates running on the party ticket. Aldrich, writing on the formation of American parties, notes that party affiliation provides candidates with a number of benefits, but none more important than the party 'brand name' (Aldrich 1995: 49). As documented in Chapter 3, United Russia has successfully developed its brand as well as its association with popular leaders, most notably Putin. This party brand provides voters with an information short cut during intensive election campaigns, representing a valuable resource for candidates.

In addition to these political benefits, United Russia also offers economic benefits as a result of its status as the preferred party agent of federal and (now) regional power, offering a means to overcome certain bureaucratic problems in the economic sphere. One problem pertains to administrative barriers, or those rules established by legislative and executive organs of power that the business community are legally obliged to follow (Auzan and Kryuchkovoi 2002: 7). A good example of the problems posed by administrative barriers can be seen in the complex procedures needed to start a new business. Before any business can open, it must first obtain permission from several state agencies (sanitation, fire inspector, etc.), which is a complicated and time-consuming task, potentially taking several years to complete. Joining a well-connected party like United Russia offers an avenue to bypass many of these barriers by giving party supporters contacts with individuals working in those state agencies charged with regulating economic activity.

United Russia also represents an interface for many economic organisations to press their interests. As noted in the previous chapter, the ability of business interests to lobby State Duma deputies has been significantly curtailed in the post-Yeltsin period. Despite this fact, affiliation with United Russia has other advantages. United Russia is the majority party not only in the State Duma, but also in most regional Dumas too, where lobbying regional laws is a less regulated affair. The party, as a collection of influential economic and political figures, represents a pool of ready-made contacts across the Russian Federation. For example, the business organisation, Delovaya Rossiya, which has close contacts with United Russia, operates as a club of entrepreneurs, using United Russia as a network of businessmen in the regions. In the uncertain economic conditions that still prevail in Russia, informal networks of reliable associates who can recommend colleagues as business partners provide opportunities for entrepreneurs who lack their own regional networks.

In addition to providing contacts that can further business interests, United Russia may also provide contacts that can protect existing business from criminal groups, as well as rent-seeking elements within the state bureaucracy itself (see Satarov 2006 for details on the latter). In this sense, United Russia resembles a *krysha* (literally a roof, but idiomatically meaning 'protection') to safeguard business interests. One of the more acute problems faced by business owners in recent years is that of *reidery*, in which businesses are physically occupied and appropriated by criminal gangs, who eject the owners by force and forge documentation to claim ownership. The problem is exacerbated by the fact that court decisions determining the actual ownership of these companies may take several years to conclude. As if to underline the seriousness of this problem, President Medvedev introduced a bill to the State Duma in April 2010 to change the legal code to combat the growing menace of *reidery*.

Joining United Russia or some of the associations under the patronage of the party, such as Delovaya Rossiya, offers the possibility of protection in the face of this and similar threats. Georgii Satarov underlines the attractiveness of United Russia as a means of protecting business interests:

> *Reidery* are strong structures – they buy courts, they buy the anti-monopoly service, they buy all the bureaucratic agencies. Very big money is invested [by these criminal groups]. So if you want your own business, the best thing to do is to run it as a representative of some power-structure or to make deals with them [representatives of power-structures], unofficial deals, of course.
>
> (Satarov 2007)

What this really means is that the biggest selective benefit of joining the party is the official sanction of the regime, in whatever form it may take. In return for showing the flag of loyalty or allegiance by joining United Russia or any other licensed party agent, elites receive something akin to a form of accreditation, which may give them certain privileges, protection or just the ability to carry on their economic and political business as usual. As previously mentioned, there are certain bounds and limits to what elites can and cannot do, but this is the whole idea of a power vertical, to create a level of responsibility among a fragmented elite that needs to submit to a higher authority. United Russia deputy, Aleksandr Krutov, communicates this need for a strong central power:

> In our country it is a rare case when a job is done properly, if there's no boss at the office. So, in our country the state plays a very peculiar role. In fact, it's necessary to have a boss not just in the office, but in every room where people work – even liberal people need a boss because they need to blame somebody. If there are no bosses, liberal people will gnaw each other – they need some common target for hatred.
>
> (Krutov 2007)

United Russia is the agent of this authority, and so has the power to attract individuals who need to get on in their career and who are willing to acknowledge a higher authority in return for some guarantees about their own place in the emerging political order.

It is obvious that United Russia provides professional politicians with (currently) the best career opportunities and so its integrative potential is undoubtedly increased by this reality. However, there is a fine line between selective benefits and soft coercion. The selective benefits that attract supporters also provide the means to control them, with the knowledge that expulsion from the party may create difficulties for their future careers, as one United Russia deputy acknowledges:

> If you are building your state career and you are excluded from the party, your relations with your boss might worsen. If you're a deputy and you are excluded from the party, you will be excluded from the faction after that too. You will become an independent deputy.
>
> (Medinskii 2007)

To understand this interplay of incentives and sanctions it is important to again consider the structure of governance that is a feature of Russian political life. This structure of governance is characterised by the prevalence of informal institutions that provide power-holders with additional opportunities to influence outcomes and the course of politics. This structure of governance boosts the ability of United Russia to attract supporters, but also to control them. To consider this point it is worth looking at United Russia's supporters in larger society, beyond the electoral arena. In the first instance, many United Russia supporters who participate in political events often do so with a clear understanding that they will receive some kind of financial reward for their efforts. It has been suggested that many young people who participate in youth organisations, such as United Russia's Molodaya Gvardia, do so, not out of any deeper ideological motivation, but because they receive free phonecards and other goods in exchange for their time. In a similar way, there is no question that youth activists who take part in counter-actions are compensated for their expenses – even if these counter actions occur beyond the borders of Russia. This was the case in 2007 when representatives of several pro-Kremlin youth organisations, including Molodaya Gvardiya, travelled to Tallinn to protest at the decision of the Estonian government to remove a Soviet war memorial.[3]

Conversely, some United Russia members appear to support the party because of pressure from their employers, who in turn receive pressure from regional party branches to mobilise their workers in support of the party. Although a difficult area to investigate, evidence suggests that some United Russia volunteers who perform tasks such as delivering election campaign material are simply given time off work and told by their employers to do party work.[4] In other cases, there is a suggestion that individuals are not only seconded to do party work now and then, but also pressured to join organisations such as United Russia and Molodaya Gvardiya outright, by either regional administrations or regional branches of the party. For regional branches, increasing the number of party members is advanta-geous as it enhances the standing of the branch within the federal party structure. In the case of regional administrations, boosting the membership of organisations such as United Russia or Molodaya Gvardia demonstrates the effectiveness of the administration, as well as their continued loyalty to the federal executive branch. The role of universities, academic institutions and businesses in recruiting young people to join pro-Kremlin youth organisations and United Russia is largely unverified, although anecdotal evidence suggests that this does indeed occur, perhaps even on a large scale.

There is also a strong suggestion that the party occasionally takes a hard line with individuals who fail to show the required level of support expected of them, with evidence that party branches pressure businesses to donate money. In November 2007, prior to the State Duma election, a scandal emerged surrounding a letter from the Kemerovo branch of United Russia informing the director of a Siberian coal company that his refusal to contribute to the party's election cam-paign fund was 'a refusal to support Vladimir Putin and his constructive course' (Lavrenkov and Romanov 2007). The letter, which made its way onto the internet, then went on to threaten the individual in question, stating that the party would 'inform the presidential administration' of the problem. Although the letter was typed on official party-headed paper, bearing the signature of the party's regional secretary, the party moved quickly to deny its authenticity, blaming the episode on a computer virus and reiterating that United Russia asks only for voluntary contributions from the business community.

Unfortunately, in the present political climate, anecdotal evidence of soft coercion is not uncommon and so reports of pressuring members to join United Russia and other pro-regime organisations will likely continue. The important point is not so much whether party supporters really want to be supporters, but that beyond the party's obvious collective and selective benefits the structure of governance greatly increases the scope of the party of power to reach, control and thus integrate them.

The limits of integration

The previous sections have shown how the creation of the all-national United Russia contributes to the agenda of the federal executive branch. In its most basic form, this agenda is the creation of stability and, to this end, United Russia serves as a useful, extra layer of organisation or administration. However, as shown in the final part of this chapter, the origins of United Russia continue to influence its development and so the integrative potential of the party of power has its definite limits. As a top-down creation designed to overcome the negative effects of decentralisation and pluralism in the Yeltsin period, United Russia is not without its problems.

Maintaining a weak party monopoly

By this point, the strength of United Russia's integrative role vis-à-vis its electioneering and governing roles should be clear. As such, the party contributes to dominant-power politics by incorporating regional administrations and elites into a controlled but relatively weak structure. The top-down origins of the party help sustain close connections with state agencies, notably state bureaucracy, and with power in general, and this serves as an incentive for many to join the party. The party is a source of useful contacts, career openings and avenues for overcoming bureaucratic barriers and a means to show loyalty to the regime. However, this nexus between state and party also promotes corruption that, if taken to extremes, can make the party a force for disintegration. This is particularly true of electoral fraud, especially if United Russia acts in an overzealous manner when trying to maximise its vote share.

In fact, the nexus between the state and the party and the opportunities for electoral fraud have already caused problems for the federal executive branch. There are many examples of regional branches of the party 'over-fulfilling the plan' (Jack 2004: 326) and using administrative resources to such an extent that the electoral process itself appears to be undermined. This, in turn, has the potential to delegitimize election victories in the eyes of the public and to create dissatisfaction with the regime. Elections results, such as the 99.36 per cent polled by United Russia in Chechnya in December 2007, give a whole new meaning to the idea of super-majority and simply represent an example of a mismanaged election.

When it comes to elections, often it is the fractions that make a difference, not just the inflated margins of victory. One of the better known examples is the failure of Yabloko to pass the 5 per cent barrier in the 2003 State Duma election due to United Russia pushing too hard to increase its vote share, as journalist Vladlen Maximov recalls:

Yavlinskii told us that in 2003, after midnight of the day of the election, when the votes were being counted, that he received a phone call and that they [the Kremlin] congratulated him on his victory. But in the morning it turned out that the party hadn't made it past the 5 per cent barrier. According to my information from certain sources, I know that they didn't plan to kick Yabloko out of parliament. Yabloko had its place in parliament because it was necessary to have a tiny little democratic faction so those people from London wouldn't have another opportunity to throw accusations of authoritarianism at us in Russia. Yabloko didn't get seats at the Duma just by chance. They tried to raise the percentage for United Russia too hard and, as a result, the law of mathematics diminished the percentage for Yabloko.

(Maximov 2007)

The consequences of overfulfilling the plan can be far-reaching and the quote at the beginning of this chapter concerning the effectiveness of the power vertical may well be applied to elections and corruption in general. It is worth noting that electoral fraud in favour of United Russia in the Moscow municipal election, documented in Chapter 3, was allegedly connected not with elections per se but with a desire among some elements associated with the party to control the development of lucrative real estate in a sought-after region of the capital.

The problem of corruption and electoral fraud is not a unique one for United Russia – both have featured consistently in examples of twentieth-century party dominance elsewhere. What is different in the Russian case is that, although modern media technologies greatly facilitate the rapid creation of a twenty-first-century dominant party by galvanising public opinion, the same media technologies have the power to rapidly deconstruct this basis of durable dominance. Of all the allegations of corruption and fraud involving United Russia in the period 2002–10, many have come to light only because of the possibilities that modern media technologies offer. Although it serves no real purpose to document (the many) allegations against the party, one example from the summer of 2010 illustrates the problem of maintaining dominant-party and dominant-power politics in the post-industrial information age. In August 2010, the party website published photographs of activists busy fighting smouldering forest fires in Voronezh; what may be considered a routine party PR procedure. However, after some sharp-eyed internet users noticed that the smoke in the pictures had been added with the help of 'photoshop' (Kotova 2010), the story was quickly reproduced on numerous websites. One incident is, by itself, relatively insignificant, but there is a tipping point when these and similar actions begin to generate momentum. It is of no real surprise that opposition parties have started to document United Russia's violations on their websites or that United Russia has started to make counter-claims of opposition party violations in what is an evolving virtual PR war between United Russia and regime opponents.

The important factor is, of course, popular opinion and how the party's actions are perceived by the public. In terms of electoral fraud, this is a difficult area to substantiate, although a survey conducted by FOM in 2007 is quite insightful on the electoral process in the post-Yeltsin period in general. According to this survey, 34 per cent of respondents thought that parliamentary elections held in Russia over the previous 15 years were neither free nor fair and 6 per cent of respondents were personally aware of electoral violations in the post-Yeltsin period (FOM 2007c). This figure should not be a cause for alarm among the authorities, but the nature of the violations may be. The violations mentioned included bribing voters ('they gave us money, and our whole neighbourhood drank for a week'); forced voting ('they demanded that you sign a pledge to vote for a certain candidate, otherwise they would fire you'); campaigning inside the polling station as voting was taking place ('they told people who to vote for'); casting votes for people not present or even dead ('only five people on our street voted, but the records say everyone voted, including me', 'somebody cast a ballot for my parents – and they're dead'); and ballot rigging ('I gave a ride to a member of the audit chamber, and I know how they counted the votes'). As noted in Chapters 1 and 3, in procedural terms, Russian elections fall short of democratic standards and the presence of United Russia exacerbates this situation.

In addition to the potentially disintegrative tendencies associated with fraud and corruption, the nature of United Russia as a top-down party emanating from power contributes to the party's weak roots in society, which limits the ability of the party to integrate groups in wider society and not just elite groups. This, one may surmise, is largely a question of time and of the party consolidating and extending its position in society, but by the end of 2010 the party's lack of real power and even relevance places definite limits on the kind of career advancement that United Russia is able to offer even to these elite groups.

The party's overall relationship with power was detailed in the previous chapter, but to further illustrate this important point it is worth noting that the status of many party leaders and functionaries is surprisingly low in comparison to non-party elites. For example, in May/April 2003, *Nezavisimaya gazeta* surveyed a number of experts from 13 regions to estimate the influence of 100 elites on the internal and external politics of the Russian Federation (Kostyukov 2003). Despite the fact that United Russia was preparing for a State Duma election campaign in December 2003, no party representative appeared in the top 10 of the country's most influential figures, and only one, the then United Russia leader, Boris Gryzlov, was in the top 20.

By April 2007, again just months away from another State Duma election campaign and coinciding with the fifth anniversary of the Founding Congress of the party, a similar survey of elite opinion conducted by *Ekspert* magazine showed that United Russia was by no means dominating the political landscape. In particular, the category of the most influential figures in the process of decision-making in the Russian Federation yielded only five party representatives in the top 20 (Gurova and Polunin 2007: 33). What is interesting is that eight of the top 10 positions were filled by active members of the federal executive branch and that former mayor of Moscow and co-chair of United Russia's Higher Council, Yurii Luzhkov, was actually rated a more influential figure than party leader, Boris Gryzlov.

By the end of 2010, a similar pattern of influence (or lack of influence) can be seen with the principal party leaders and the party positions they occupy. According to the annual *Nezavisimaya gazeta* survey of Russia's top 100 politicians, Higher Council chair, Boris Gryzlov, and Higher Council co-chair, Sergei Shoigu, were rated as Russia's thirteenth and twenty-fifth most influential politicians respectively (Orlov 2011). One of the other senior party figures, Mintimer Shaimiev, dropped out the top 100 altogether after stepping down as president of Tatarstan, despite retaining his post as Higher Council co-chair.

The issue of the power and prestige of United Russia, as well as the overall quality of the party cadre, were called into question by the developments surrounding President Medvedev's initiative to create a cadre reserve selected by the presidential administration as a pool of future leaders. The list containing the first 100 of the presidential cadre reserve published on 1 March 2009 contained only two high-profile representatives of United Russia. In fact, United Russia decided to create its own parallel cadre reserve, selecting 1,000 promising individuals earmarked for fast tracking to senior positions. The original plan, according to United Russia, was that around 300 of the party's cadre reserve would find their way onto the presidential cadre reserve, but in March 2009 a presidential aide dismissed the idea that there would be any party quota (Nagornykh 2009b).

This development, of Medvedev deciding against using United Russia as a cadre reserve in its own right, is not only an unusual move, but also another indication of the ambiguous relationship between party and power discussed in the previous chapter. What is interesting is that, even when United Russia created its own parallel cadre reserve, there were inevitably problems with corruption, with one individual removed from the list after allegations were made that party representatives in the Ural territorial district sold him his place for 10,000 US dollars (Novye Region 2009).

Overall, this issue of party prestige is as much a consequence of the rapid party-building exercise that characterised the early years, 2002–3, as of the structure of governance and relationship between the party and other components of the power vertical, such as state bureaucracy. In the early years of United Russia's formation it was necessary to literally find party candidates and members, often at short notice, in order to staff the thousands of party offices that appeared within months of the December 2001 Founding Congress. From this perspective, it is not so surprising that a poor quality, even corrupt, element infiltrated the party and this was reflected in the campaign initiated in September 2008 to clean up the party ranks and introduce tighter procedures for party membership. However, the origins of a party designed as a stabilising force bring an inevitable amount of inertia and so elite circulation within the party is as much an issue as elite circulation in the wider political system.

Evidence for this can be seen in the composition of the United Russia faction in the Fourth Duma Convocation, just prior to the 2007 State Duma election. The average age of the United Russia faction (based on data for 308 United Russia deputies) was 52, which meant that most deputies were old enough to be CPSU members. In fact, 66 per cent of the faction (205 out of 308) stated their previous

CPSU party affiliation. By December 2010, the average age of the United Russia faction in the Fifth Duma Convocation was 53 years of age. To put this in perspective, the average age of the CPRF faction – the 'party of the past' – was 58 years.[5] There is a case to be made that the party of power from the late Soviet period to 2010 is, in fact, the same party, comprising more or less the same, former Soviet, elite.

Despite the party's commitment to provide more opportunities for women and young people to enter politics, so far this remains a problem for the party leadership. Although over 60 per cent of party members are female, the United Russia faction in the Fourth Duma Convocation contained only 22 female deputies (7.1 per cent of the faction). Although the Fifth Duma Convocation (2008–11) saw the number of female deputies more than double to 45, this was still only 14 per cent of the total party faction. In 2006, United Russia promised a 20 per cent quota for young people on its party-lists, from organisations such as Molodaya Gvardiya, in an effort to boost the number of young people entering politics. However, the party-list for the 2007 State Duma showed that under 35-year-olds comprised 11 per cent of the total, with only 11 activists from pro-Kremlin youth movements represented among the 600 United Russia candidates competing for 450 mandates in the State Duma (Yashin 2007). The party faction in the Fifth Duma Convocation, as of January 2008, contained 32 deputies who were 35 years or younger, a little over 10 per cent of the total faction (Duma 2010b). A deputy in the United Russia faction of the Moscow Legislative Assembly comments on career opportunities for young people within the party: 'we have problems creating an atmosphere of social mobility for them. They join the party, but can't see their career prospects after their joining. Some of them come and then leave the party, some wait for their opportunities' (Moskvin-Tarkhanov 2007).

Despite the fact that the party federal leadership and the federal executive branch understand the importance of elite circulation within the party, initiatives such as the 20 per cent youth quota often fail to generate enthusiasm among regional party branches when they come to draw up their candidate lists.

So far there is no serious indication that elites are unhappy with the limitations on career advancement within United Russia, or, more generally, in the centralised post-Yeltsin system. However, the appearance of A Just Russia in late 2006 may be broadly interpreted as an acknowledgement on the part of the federal executive branch that more needs to be done to provide for elite circulation. It was mentioned in Chapter 3 that A Just Russia provided extra party positions as a consequence of the abolition of SMD voting from 2007, but also for those elites unable to satisfy their leadership ambitions in United Russia. Gennadii Gudkov, a former leader of the People's Deputies deputy group and member of the United Russia faction up until the spring of 2007, commented on his subsequent defection to A Just Russia:

> Putin understands that the absence of polemics or political struggle is bad for the state and that's why they made this project, A Just Russia, to revive political life. Of course, A Just Russia is not real opposition, but it is an acceptable

level for Putin and the Kremlin. It gives some possibilities to politicians like me, some political freedom . . . we have more possibilities for our thoughts and ideas, but they have to give us more opportunities.

(Gudkov 2007)

A Just Russia is by no means a solution to the general lack of power and prestige of political parties in Russia. At the same time, the constant production of licensed opposition parties is a short-term solution to the limits of elite circulation in the political system, but also within United Russia itself.

A solution to the problem of collective action?

Another consequence of the way that United Russia was formed is the fact that the party is home to a diverse range of opinions. This is understandable in view of not only the rapid expansion of the party of power, but also its purpose to integrate elites from across the Russian Federation into a structure controlled by the federal centre. However, United Russia is not an organisation designed to attract part of the elite – those individuals sharing similar concerns and beliefs about Russia and its political and economic development. Instead, the party is an attempt to place some control over the elite as a whole. Former Union of Right Forces leader, Nikita Belykh, summarises this point well: 'There are very different people in this political party . . . United Russia is a constellation of people who are connected by a common interest, which is to keep their positions in the power hierarchy and to maintain their businesses' (Belykh 2007).

Indications of the potential for these diverse opinions to lead to conflict can be seen in the party's early efforts to formulate an official ideology in 2005/6, resulting in the emergence of distinct intra-party platforms or wings. In fact, two of the party's influential internal 'clubs', the Club 4 November and Centre for Social Conservative Policy, were created in 2005. The intra-party platforms reflected the clear divisions between liberal groups in the party – with figures such as Vladimir Pekhtin (deputy head of the party faction in the State Duma), Vladimir Pligin (General Council) and Vladislav Reznik (Presidium of the General Council) broadly reflecting the interests found in the Club 4 November – and conservative and *siloviki* elements represented by the Centre for Social Conservative Policy, headed by Yurii Shuvalov and considered close to the ideas of Boris Gryzlov (chair of the Higher Council), Sergei Shoigu (co-chair of the Higher Council), Andrei Vorob'ev (Central Executive Committee) and Yurii Volkov (General Council), among others.

Comparative experience, in particular that of the Japanese LDP, shows that distinct ideological platforms do not necessarily undermine the integrity of the party, but can have a positive effect if conflict is controlled within limits. In the Russian case, the experiment with party 'wings' was quickly sidelined when it became clear that levels of conflict between competing ideological interests posed a risk of escalation beyond the confines of the party. Ideological conflict may not have been an issue among the higher echelons of the party, but there was the possibility that, in the regions, ideological platforms may have exacerbated existing levels of conflict within the party, among competing economic groups.

In any case, the liberal/conservative wings within the party have not disappeared, even with the adoption of Russian conservatism as the official party ideology in November 2009. In December 2010, one of the leading figures of the liberal platform and one of the senior figures in the party as a whole, Vladimir Piligin, made the surprising prognosis that the party may need to be reformed, because of divergent opinions in the party leadership on questions of party ideology and its future role (Rodin 2010).

These internal disputes lend some credence to the idea that United Russia is not a genuine party-based solution to the problem of collective action, but a forced solution imposed on terms that strongly favour the conservative, statist element in the ascendency since 2000. However, the ability of the federal executive branch and United Russia to actually impose this solution on every region consistently is open to question. As already noted, in formal terms, United Russia is highly centralised and this organisational feature is useful for maintaining the integrity of the party, but also for enabling a coherent command structure across a large territory. However, there is some uncertainly as to the actual degree of centralisation within United Russia. This situation – of a party being less centralised than it claims – appears to be the reverse of usual comparative experience. Duverger, writing on parties in the early twentieth century, noted that many parties claim to be decentralised even when they are, in fact, highly centralised. This, Duverger notes, is because party centralisation has a 'pejorative flavour' (Duverger 1964: 56). In the circumstances of the post-Yeltsin period, in which centralisation symbolises elite consensus and unity, the reverse seems to be true.

In both the Yeltsin and post-Yeltsin periods, party of power party-building has been characterised by territorial penetration rather than territorial diffusion; that is, the (federal) centre has controlled and directed the development of the party in the peripheries (Panebianco 1988: 50). However, the success of party of power penetration into the regions has been largely dependent on the support of regional governors and administrations. Although the relationship between the party of power and regional administrations varies from region to region and so is impossible to generalise, there is some support for the idea that party central office simply coordinates and manages pre-existing regional electoral networks, supplying them with the United Russia label and allowing regional elites to develop the party franchise as they see fit. So far, one of the clearest indications that this is indeed the case (and perhaps a worrying development for the party and regime) was the decision made by the Tuva branch of United Russia in August 2010 not to use campaign slogans associated with either the party or Vladimir Putin in the October regional elections (*Vedomosti* 2010). The Tuva party branch considered 'United Russia' and 'Putin's team' unpopular messages for the region's voters.

This somewhat contradictory organisational arrangement resembles Eldersveld's (1964) elaboration of 'stratarchy', a dilution of power throughout the party structure. Eldersveld states that this allocation of command and control is important when the 'heterogeneity of membership, and the sub-coalitional system make centralised control not only difficult, but unwise' (Eldersveld 1964: 9). From this perspective it appears far from certain that United Russia has managed to successfully penetrate every region and to create an unmediated party structure that effectively supplants the authority of pre-existing regional elite groupings. Former United Russia (2004–7) and current A just Russia State Duma deputy (2008–11), Mikhail Emel'yanov, makes this candid observation on the relationship between United Russia's election results and regional administrations:

> A favourable atmosphere has been created for the party. No doubt, local authorities create this atmosphere for United Russia. If we take into account the fact that many members of the local authorities are United Russia members . . . it's possible to say that the achievements of the local authorities are the achievements of United Russia too.
>
> (Emel'yanov 2007)

In fact, confirmation of this arrangement comes not from United Russia, but from the federal executive branch itself. In 2006, Vladislav Surkov noted that, in many regions, it is the party that relies on the regional administration rather than the regional administration that relies on the ruling party:

> In the overwhelming majority of regions, United Russia relies on the incumbent authorities – regional leaders, city mayors, and so on. I don't want to go into an analysis of why this situation has come about, or how stable it is. United Russia's task is to make an orderly transition from relying on administrative props to keeping afloat by itself.
>
> (Surkov 2006b)

As noted in the opening section of this chapter, the official rationale for the creation of United Russia lies in the threat of territorial disintegration, which, according to recent experience with Chechnya, is most likely to occur in the national republics. For this reason, one may view President Medvedev's recent gubernatorial appointments as evidence that the federal executive branch is seeking to weaken the independence of regional administrations, which may serve to strengthen the hand of the party in the future. To illustrate the extent of recent governor appointments it is worth noting that, in May 2008, when Medvedev began his first term as president, it was possible to count several chief executives in the national republics who had held power for most of the post-Soviet period – Bashkortostan (1993), Kalmykiya (1993), Mordoviya (1995) Tatarstan (1991), Chuvashiya (1993) – or who had held power for most of the post-Yeltsin period – Karelia (2002), Komi (2001), Marii-El (2001), Sakha (2002), Udmurt (2000).[6] By January 2011, and with the exception of Mordoviya, Marii-El and Udmurt, Medvedev has replaced regional leaders in each of these republics.

Of course, every region is different, and the ability of United Russia to superimpose itself over existing administrations and electoral machines to create, in effect, one huge, all-national electoral machine is a hallmark of the party's success in the post-Yeltsin period. From the perspective of the federal executive branch, it remains to be seen if short-term and long-term stability is better served by such a flexible approach to the national republics and, indeed, to other regions in the Federation. In fact, the flexible approach of the Yeltsin administration, of creating bilateral agreements with certain regions, had its advantages but nonetheless counters the official rhetoric, even rationale, of the power vertical, United Russia and the post-Yeltsin regime itself, to curb the centrifugal tendencies that these agreements were deemed to encourage.

What is interesting is that both the federal executive branch and United Russia were instrumental in renewing the bilateral agreement with Tatarstan in February 2007, despite strong opposition from many political figures, including those within United Russia. During a Duma debate on this agreement on 9 February 2007, LDPR deputy, Aleksei Mitrofanov, had his microphone switched off in the middle of his speech criticising the initiative (Duma 2007). Independent deputy in the Fourth Duma Convocation, Viktor Alksnis, a strong opponent of separate deals with the regions, recalls how the State Duma approved the agreement with Tatarstan in February 2007:

> The agreement with Tatarstan was due to be discussed on Monday, at 4 PM. However, at the very last moment, United Russia took the decision to change the time – they announced that the issue would be discussed at 3 PM instead of 4 PM. Their plan was simple – not all the deputies got the information about this change of time and were not around during the break. So it was easy for United Russia to push the approval in an almost empty room full with mostly United Russia deputies.
>
> (Alksnis 2007)

In this respect, United Russia sometimes appears to be something different from the party charged with integrating Russia's regions, borne out of the fight against the tendencies of regionalism that existed at the end of the Yeltsin period.

Concluding remarks

United Russia was built to integrate elites as part of a larger effort to rationalise centre–periphery relations. The party provides an extra layer of administration across the regions and has a significant pull in attracting supporters. United Russia is a factor in the persistence of dominant-power politics because it corrals supporters and potential opposition into a controlled structure. Although United Russia cannot be considered an explanatory variable in the rise of dominant-power politics, it is a factor in its persistence, seen clearly through its integrative role. As the final section demonstrated, the party does have its limitations as an integrative force, reflecting the origins of United Russia, as well as the larger weakness of a party created as an agent of power.

6 Parties and regime outcomes
The post-Yeltsin ruling party

Opposition depicts us [United Russia] as the party close to government and talks of us as if we represent all the power in the country. But in reality, we don't represent all the power in the country. I have a very different opinion in this regard. I believe that we have been preparing the party to become a real party of power.

(Viktor Grishin 2007)

It's not a party. How to make real parties, nobody knows.

(Sergei Markov 2007)

The previous chapters have provided a detailed theoretical and empirical discussion of parties and their relationship with larger, macro-political outcomes, using the case of United Russia and dominant-power politics in the post-Yeltsin period. After discussing the nature of dominant-power politics in Russia, 2000–10, and presenting a review of the typical mechanisms through which ruling parties generate this outcome, Chapters 2–5 then showed the exceptionalism of the Russian case, how context and history combine to reverse the arrows of causality, making the party an outcome of dominant-power politics rather than its cause.

This concluding chapter now returns to the two questions posed at the beginning of this study, before broadening the inquiry to consider three related themes. The first theme is the nature of United Russia as a type of party and what, if anything, the previous chapters tell us of the party of power in comparative perspective. The second theme is also comparative in scope and concerns the contribution that this case study makes to existing literature on party politics, as well as the 'new authoritarianism' of the post-Cold War period. The third theme switches attention back to the Russian case and considers the prospects for United Russia and the emerging post-Yeltsin order in the future.

Dominant-power politics and party roles

The introduction chapter presented two questions or foci that formed the basis of this study. They included the nature of United Russia's relationship with dominant-power politics and the party's overall role in the politics of the post-Yeltsin period. The first point that becomes immediately apparent from the previous chapters is that United Russia does contribute to the regime, broadly in line with the theoretical framework presented in Chapter 1. Each chapter showed that United Russia has some input into managing elections, governing and enforcing incumbent authority and integrating elites and society. In more detail, and building on the previous chapters, it is possible to say that United Russia rationalises the party-system by concentrating regime support into one party, while restricting opposition parties in the formal institutions of governance (Chapter 3). The party also rationalises executive–legislative relations by providing stable, pro-government and pro-president support in the State Duma, to ease the passage of legislation and limit the effects of veto players (Chapter 4). The party also rationalises centre–periphery relations by corralling elites from across the regions and subjecting them to the discipline of the party. The question of the role of United Russia in the post-Yeltsin period is thus answered: the rationalisation of politics by extending control of the federal executive branch over the party-system, state legislature and the regions.

However, the previous chapters have also developed a strong counter argument that United Russia is not a typical ruling party and so diverges quite markedly from the ruling party ideal type outlined in Chapter 1. Understood in terms of the principal–agent relationship, United Russia is an agent of non-party power-holders residing in and around the federal executive branch, rather than a principal power in the political system. This relationship appears to be the reverse of that typically, but not always, found in comparative literature in which ruling parties are central in their respective political systems and, consequently, in a range of regime outcomes. United Russia did not create the post-Yeltsin regime and, by the end of 2010, it remains a component, but not source, of dominant-power politics in Russia.

Chapter 5 showed that United Russia's ability to manage elections in the post-Yeltsin period is significantly enhanced by pre-existing incumbent strength, that is, existing control over primary resources, such as financial flows, informal institutions, media outlets and, in the case of Vladimir Putin, and to a lesser extent Dmitrii Medvdev, high levels of personal popularity among voters. Clearly, the popularity of regime leaders is itself an outcome of these other resources, but the point is they all are non-party resources, controlled directly by either the federal executive branch or regional executives, and are then discretionally allocated to party agents. To date, United Russia has been the net party recipient of these primary resources and other secondary resources, as the regime's preferred party agent. But it is important to reiterate that managing elections is a multi-party strategy, in which regime parties and so-called opposition parties work to further the aims of the federal executive branch. These aims can be identified (in formal terms, at least) as securing political stability and economic development, together with a reaffirmation of the territorial integrity of the Russian Federation and the pursuit of a national interest defined by the federal executive branch. Democracy also remains a central component of the official state discourse; even if its most significant elements, such as electoral competition, have been significantly watered down.

Chapter 4 showed that United Russia's status as a governing or ruling party is open to question, especially in view of the fact that federal (and most regional) governments continue to be formed along professional or non-party lines. This situation is gradually changing in favour of United Russia, but there remains enough ambiguity surrounding the relationship between party and power-holders to say that, at the current speed of development, Russia is some way off the emergence of a ruling party in any meaningful sense of the term. Despite Vladimir Putin taking up the party leadership in 2008, he is still not a party member, so the degree of control that the party has over the federal executive branch remains negligible. Although it is indisputable that senior figures within United Russia form part of Russia's ruling clique, they are no more the agents of the party than the president or prime minister. In general, the party has little input in the formal institutions of power or in enforcing incumbent authority outside of them.

Chapter 5 showed the party to have a much more positive relationship with dominant-power politics, clearly an area in which the party plays an important role in maintaining the status quo. In practical terms, United Russia has enjoyed some success in compensating for the absence of a comparable structure for elite socialisation during the Yeltsin period. However, United Russia's lack of real power in the political system, its lack of ideology, the variety of views within the party and the lack of elite circulation, within the party and the centralised political system, places limits on this integrative role. In short, and as shown in Chapter 5, regional elites often join United Russia to show their loyalty to the federal executive branch and a case could be made that United Russia still offers few exclusive party resources that are not available to regional elites elsewhere.

As already stated, neither United Russia nor Unity created the post-Yeltsin political order, but United Russia does appear to play a role in maintaining it. In addition, the overall legitimacy that the party confers on the regime should not be discounted, and in many ways this is one of the biggest hidden contributions that United Russia makes to the persistence of dominant-power politics. The existence of parties and a working legislature make claims of Russian democracy plausible, even if both party-system and parliament are tightly controlled. However, United Russia is by no means a strong explanatory variable, often appearing as an outcome of other processes, a result of the intentions and capacity of power-holders to project their strength on to their preferred party agent(s). At best, the relationship between United Russia and dominant-power politics is one of reciprocal causation, as noted in Chapter 4 with the process of centralisation, in which increases in the strength of one contributes to increases in the other. United Russia may also be understood as an important intervening variable in the Russian political system, one of several tools used by the federal executive branch to consolidate power.

Why not an explanatory variable?

To understand why United Russia is not an explanatory variable for dominant-power politics in the post-Yeltsin period it is necessary to look at context, at party origins and at the larger circumstances in which the Russian state emerged from the collapsing Soviet Union in 1991. As discussed in Chapter 2, the Russian Federation emerged during a process of de-partification, not only democratisation – a transition away from single-party rule – and this study has shown that, even by the end of the second decade of post-Soviet politics, evidence of this earlier de-partification is very tangible. As noted in Chapter 4, there are still laws restricting the activities of parties and, even though restrictions on government ministers holding leadership positions within parties have been removed, Russia's rulers have so far resisted institutionalising their power within any party organisation. This, as discussed in the material that follows, has implications for United Russia, but also for the stability and future political development of Russia as a whole.

The weak position of parties in the Russian Federation is also attributable to the fact that Russia's earliest post-Soviet power-holders, at both federal and regional levels, did not owe their position to any political party. Parties, although proliferating from the late Soviet period onwards, were not central to the emerging political system. The Constitution of the Russian Federation, adopted in December 1993, reflected this reality, creating the conditions for a powerful presidency and basis for future incumbent capacity that was not reliant on parties. By the 1995/6 electoral cycle, the federal executive branch had already dispensed with the fledgling, grass-roots democratic movements and created the first completely top-down parties of power, strictly subordinated to the federal executive branch – in the case of Our Home is Russia, a prototype for the later United Russia.

These factors mean that post-Soviet and post-Yeltsin Russia continues to have relatively low levels of what may be termed 'partyness', of positive public and elite estimations of parties and high levels of party power and prestige within the political system. Continuing low levels of partyness mean that parties in the Russian Federation tend to be either agents of power-holders, as is the case with parties of power, or, conversely, agents of party leaders – arguably the case with most opposition parties that, as noted, show a curious continuity in their leadership. In view of this reality, the use of terms such as 'hegemonic', 'dominant' or even 'ruling' party to describe United Russia is misleading, although understandable in view of the odd, even deliberate, mismatch between its physical form, party-system dominance and image and its lack of power and peripheral nature.

In fact, the first contribution that this study makes, in comparative terms, is to restate an older, but often overlooked, clarification in the literature on the subject of party dominance and hegemony. Although parties that dominate their respective party-system, but not their political system, are rare occurrences, this case study serves as a reminder that these parties do nonetheless exist. As stated in Chapter 1, party dominance is usually measured in terms of party electoral returns, consecutive election victories and seats held in parliament (Magaloni 2006: 36–7; Pempel 1990b: 3–4; Sartori 1976: 174), but there is an implicit power dimension too, seen in terms of party control over government, appointments, policy-making and decision-making in general. Reddy captures this power dimension succinctly with the comment that 'the dominant party is the only arena that really counts; all policy action takes place within it' (Reddy 2006: 59). This is certainly not the case with United Russia, as the previous chapters demonstrated.

In a similar way, Sartori's seminal contribution to the subject of party dominance also carries an implicit, but often understated, power dimension, which
is worth considering in more detail. According to Sartori, party dominance in
democratic systems is characterised not by a dominant party, but by a predominant party-system. In contrast, party dominance in non-democratic systems is
characterised either by a single party (as is the case with the CPSU in the Soviet
Union), where only one party 'exists and is allowed to exist' (Sartori 1976: 197),
or by a hegemonic party, which 'tolerates and discretionally allocates a fraction
of its power to subordinate political groups' (Sartori 1976: 205). The hegemonic
party tolerates 'licensed opposition', but in the words of Sartori the 'out parties'
can never become 'in parties', reflecting the lack of real competition in these
political systems.

The hegemonic party type is revealing because it shows a clear hierarchy
of power within the political system as a whole. The hegemonic party rules or
governs at one level, while licensed opposition parties compete as a façade at
a second level; the latter are never permitted to hold real power in the political system. Extending from this, one may surmise that the dominant party in a
democratic system operates in conditions in which all parties are de facto and
de jure permitted to compete and to rule, just that one party is particularly successful in outperforming its party rivals. In contrast, in the single-party state, the
ruling party holds power in totality, without competition and rivals, operating in a
political system with no formal competition, only informal competition confined
within the single party.

In post-Soviet Russia, political power at the highest level is still not organised
along party lines, despite the gradual rise of United Russia. As such, Sartori's
elaboration of party dominance and party hegemony seems largely irrelevant for
the Russian case. As of 2010, the Russian political system may be said to have
several centres of power or principals in the political system, including the federal executive branch, regional administrations, state bureaucracy, the security
services and other 'shadow' power-holders, but United Russia and other political
parties are merely appendages or agents of power. Although the nature of the
post-Yeltsin regime remains underelaborated (Sakwa 2010: 32), it is nonetheless
possible to talk of a dominant regime, although not a dominant-party regime.

United Russia in comparative perspective

Although this study has clarified both the relationship of United Russia with
dominant-power politics and the role of the post-Yeltsin party of power, there
still remains the question of how to understand this party in more comparative
terms. Is it possible to identify a type of party that has comparative relevance for
the Russian case, and so facilitate future comparative research? So far, elaborat-
ing the party of power as a type of party has proved difficult, in no small way
reflecting the deficiencies of the term already identified in previous chapters.
Oversloot and Verheul (2000: 136) observe that the party of power 'does not
fit well with known concepts for categorising political parties' and, as already
stated, the party of power appears to be a phenomenon that is simultaneously

both more and less than a typical political party, understood in classic terms at least.

In general, the comments made by Oversloot and Verheul correspond with those made in Chapter 2 that 'party of power' is not an analytical term, but a political term that captures a general reality, but one that is useful only for distinguishing pro-regime from anti-regime parties. Not only is the 'power' component of the party of power term problematic, but so too is the 'party' component, and this is one immediate and obvious obstacle to fitting them into comparative categories. From the emergence of Our Home is Russia in 1995, the party of power resembled a bureau rather than a party, something akin to a 'department for voting and elections' controlled by federal government. In more abstract terms, the etymology of the word 'party', derived from the Latin *partire* (to divide), does not sit easily with the nature of the party of power as a whole rather than a part. United Russia, like the CPSU in the Soviet period, is only party-like at the bottom, with the top often appearing indistinguishable from the state.

There is no question that, on the basis of this study alone, there are certainly grounds for dismissing the party of power as a party at all, and to reframe it in terms of state structures. In this sense, this study has indulged the considerable efforts made on the part of the post-Yeltsin regime to sell its management tool as a genuine party, with all the positive associations of elite unity, popular support and democracy that this entails. As a result of this study's overall approach of engaging with United Russia and the party of power as party organisations, five types of party have at various points been highlighted. Chapter 3, for example, identified United Russia as an example of a 'cadre party' and a 'catch-all party', while the introduction chapter mentioned 'dominant party', 'hegemonic party' and, of course, 'ruling party'. The latter three have their serious limitations and either cadre or catch-all may be a more useful comparative type.

Of course, any party type is inevitably incomplete, usually focusing on one particular aspect of the party. Catch-all party usefully acknowledges the party of power's electoral strategy as it has developed over time, while cadre party identifies some prominent organisational features, notably its reliance on funding from sources other than its membership. In fact, none of these party types is mutually exclusive, meaning that parties of power can be said to be catch-all, cadre and several other party types at the same time. However, despite the difficulty in classifying the Russian party of power, party types are more useful when they focus on the most discernible characteristics of the particular party in question.

In view of the nature of the post-Soviet Russian political system elaborated previously, neither catch-all nor cadre party highlights the most significant characteristics of the party of power. This study has persisted with the party agent label in order to draw out the relative lack of power that this party has, but beyond Sartori's aforementioned 'ambassador party' no existing party type really captures this reality adequately. This study does not advocate generating yet more party typologies, so the closest existing comparative party type applicable for the party of power is the 'personalistic' party proposed by Gunther and Diamond (2003). Gunther and Diamond define this type of party as:

An organization constructed or converted by an incumbent or aspiring national leader exclusively to advance his or her national political ambitions. Its electoral appeal is not based on any programme or ideology, but rather on the personal charisma of the leader or candidate, who is portrayed as indispensable to the resolution of the country's problems or crisis.

(Gunther and Diamond 2003: 188)

Although the personalistic type of party highlights features that have been emphasised in the previous chapters, such as the importance of the 'Putin factor' or more generally the 'elite variable' on party of power development, this category is by no means unproblematic. The party of power has an explicit and clear relationship with power-holders that is not captured by the personalistic type of party. Personalistic parties may be opposition parties, which does not fit with the general idea of the party of power as a structure 'built to support power-holders'. Gunther and Diamond also note that the organisation of personalistic parties tends to be weak, shallow and opportunistic (Gunther and Diamond 2003: 188), which, although appropriate for the Yeltsin-era parties of power, does not entirely fit the experience of United Russia in the post-Yeltsin period. However, if party types are viewed as ideal types, it is neither important nor possible that a party demonstrates all of the associated characteristics.

In terms of comparative analogies, the party of power's origins mean that United Russia combines the dependence of a personalistic party with strong bureaucratic elements. Probably the closest comparators are other parties 'dependent on power-holders'. These include parties created as civilian fronts for military regimes such as the Indonesian Sekber Golkar mentioned in Chapters 1 and 5, but also parties inside and outside the post-Soviet space created as agents for presidents or ruling cliques in and around a civilian-controlled executive. These examples are often referred to as personalistic parties and include Alberto Fujimori's quartet of Cambio 90, Nueva Mayoria, Vamos Vecino and Peru 2000 (Gunther and Diamond 2003: 188), among others.

Parties as explanatory variables

In many ways, the appearance and emergence of United Russia in the post-Yeltsin period confirms existing comparative literature on the subject of party formation and development. For example, Shefter (1994: 7) notes that mass parties emerge when incumbents are forced to counter-mobilise when faced by an opposition that has already mobilised popular support. In the Russian case, it has been suggested that United Russia emerged as a reaction to the threat presented by FAR at the time of the 1999 State Duma election, a beefed-up version of Unity to counter the threat posed by regional elites (Hale 2006: 237). Although the previous chapter dismissed the threat of FAR as a factor in the emergence of United Russia, the emphasis on human agency and the elite variable is justified.

Elsewhere and as already noted, the appearance of United Russia as an all-national party appears to confirm the work by Chhibber and Kollman (2004) on the emergence of national party-systems. According to this study, when the federal centre becomes more important in the political and economic life of the state and the nation, all-national parties become more important vis-à-vis regional parties. This is certainly true in the Russian case, where centralisation, high levels of popularity for Putin and certain restrictions placed on party formation and non-party-affiliated candidates contributed to the increased importance of the federal centre and the emergence of an all-national party-system from the de-centralised and fragmented party-system of the Yeltsin period.

This case study, together with the material presented above, also allows some comments on the value of parties as explanatory variables in similar political systems. This has become increasingly relevant in the post-Cold War period as the initial positive results of democratisation seen in many parts of the world, including the post-Soviet space, have, in many respects, petered out. Dominant-power politics or electoral authoritarianism is increasingly common and so our understanding of the way that these new forms of authoritarian regimes are built and maintained is likely to be increasingly relevant in the future.

The first contribution that this study makes comes in the form of a reminder that there is more heterogeneity among ruling and dominant parties than existing literature acknowledges. The assumption is often that, if a party appears to be dominant in the party-system, it is probably dominant in the political system too. In many, but not all, contexts this is indeed the case. The recent move towards more quantitative measures of regime type based on the degree of competi-tiveness in the political system is not necessarily helpful in this respect. These measures tell us a great deal about the democratic credentials of a particular state, but little about the way that power is organised. Of course, regime clas-sifications, like the party classifications discussed above, inevitably encounter the same problem of focus. At present, the focus is on the residual category of non-democracy, in order to separate states that are democratising from those that have halted this process somewhere along the way, in some cases not very far along the way. However, identifying the sources of incumbent strength in these regimes requires a focus on the way that power is distributed and organised in the political system in formal and informal terms, as well as on the degree of electoral competition they exhibit.

The second contribution that this study makes is to provide a more sober assessment of the relationship between parties and larger regime outcomes. This chapter has already highlighted certain contextual features that contrive to weaken the role of parties. In Russia, the reason why a strong executive branch emerged free from party constraints in the first place is in no small part attribut-able to what may be termed 'bad party history' – the legacy of one-party rule and the continuing popular distrust of parties. It is worth noting that in Peru under Fujimori (1990–2000), in a regime in which parties were also built by the execu-tive branch as top-down personalistic parties or party agents, this development emerged against a similar historical backdrop.

In Peru, prior to the election of Alberto Fujimori in 1990, political parties had become largely discredited in the eyes of politicians and voters alike. From the appearance of the first and largest mass-party in the 1930s, the Alianza Popular Revolucionaria Americana (American Popular Revolutionary Alliance), Peruvian parties embarked on a slow and gradual slide, becoming empty shells by the 1990s (Roberts 2006: 82). Political and economic mismanagement led to Peru's bad party history and their abandonment as serious organisations. Parties remained weak and ineffectual and, by the 1990s, were easily dominated by the powerful Fujimori presidency. Their very weakness contributed to authoritarianism in Peru under Fujimori by removing a potential check on his presidential power.

Often the role of parties operating in this context of bad party history is largely symbolic, with them serving only as a very demonstrable sign of democratic development for domestic and international observers. This symbolic use of parties is not a new occurrence. In an earlier period it has been noted that parties were sometimes created as little more than signs of a particular kind of modernity, status symbols for a regime or ruling elite, demonstrating their progress and political development to their own people and the rest of the world (LaPalombara and Weiner 1966: 4). In the post-Cold War period democracy is a political system without an obvious or credible alternative, so political parties remain synonymous with freedom, pluralism and constructive politics. Their inclusion in electoral authoritarian systems, such as Russia's, serves practical purposes, as this study has demonstrated, but also the important decoration necessary for the democratic façade and for the legitimacy of the regime as a whole. As the quote at the beginning of this chapter indicates, so little is known about the way that 'real' parties are built and emerge. In the Russian case, with the weakness of society and relative strength of the state, it is difficult to see where they would come from if not from the state and those who hold power. In such circumstances, these parties are likely to reflect not only the intentions of power-holders, but also their strength and the kind of resources they control.

If the party variable is not explanatory in the rise of dominant-power politics in the post-Yeltsin period and if the value of parties as explanatory variables may be overstated (in the Russian case at least), the important question is: what are the sources of dominant-power politics? Although this study did not focus directly on this question, the previous chapters have highlighted a number of areas worth exploring in more detail. The basis of incumbent strength in post-Soviet Russia is a combination of control over financial flows, the structure of governance, state control over media outlets and the strength of the president, including his own popularity.

In addition, the maintenance of dominant-power politics and more generally authoritarian regimes may again be less connected with the strength of the ruling party as it is with the presence of an effective multi-party system. The Mexican case is highlighted as an example of long-term stability created through a ruling party, but the Mexican system was historically distinguished from the Soviet system by the presence of a semi-competitive, multi-party system. This is certainly not a new occurrence and comparative experience provides plenty of examples of authoritarianism with multi-party systems. In fact, it is not difficult to find examples of electoral authoritarianism in which control over a multi-party system contributed greatly to dominant-power politics, such as in Egypt (Albrecht 2007: 61), Jordan (Lust-Okar 2007: 53) and Azerbaijan (Ottaway 2003: 95).

In the Russian case, the existence of numerous parties, some formally in opposition, others in support of the regime, contributes greatly to the stability and legitimacy of the political system. Shefter, writing on the conditions in which incumbents may choose not to build a strong mass-party, notes that this may be the case when 'politically excluded groups are divided or can be intimidated' (Shefter 1994: 9). Overall, the weakness of opposition, and the degree of control that the federal executive branch currently has over the party-system, probably contribute to the weakness of United Russia and the decision, thus far, not to institutionalise rule fully within the party.

Future political development in Russia

Ultimately, it is difficult to look too far ahead and draw conclusions about the implications of United Russia and the future development of Russian politics. As Geddes rightly observes, 'it is hard to explain an outcome that has not yet finished coming out' (Geddes 2003: 64), and in many respects United Russia and the post-Yeltsin political order are still developing and adapting to circumstances. This is really the point when it is necessary to underscore an extremely important factor that has and will be key in the third decade of post-Soviet politics – the role of contingency. However, it is possible to make some observations based on the previous chapters regarding United Russia, the party of power and the future of the post-Yeltsin regime.

First, and despite appearances to the contrary, United Russia does not indicate a party-based resolution to the problem of collective action in the Russian Federation. At best, and as mentioned, United Russia represents a forced elite settlement that strongly favours one group of ruling elites over all others. However, as shown in the previous chapter, the ability of the federal executive to force regional administrations to submit to the party is questionable, as is the wisdom of this course of action. As such, by the end of 2010, United Russia appears to be at a crossroads, as does the entire post-Yeltsin political system. As the post-Yeltsin regime is a pause between the closed Soviet political system and genuine competitive democracy, there is a constant pressure for the regime to take the next step, and this can be seen in the constant discursive output from the federal executive branch in terms of modernisation, Strategy 2020 and other messages of 'direction' and a bigger plan. What is not clear is if this means more or less regime control over political processes. In the meantime, the stability of the post-Yeltsin order will depend on the ability of incumbents in and around the federal executive branch to perpetuate, what is in many ways, a poorly institutionalised political system. The previous chapters have presented an argument that United Russia is an agent of power-holders, a reflection of their strength and intentions, rather than a source of strength in its own right. As such, the party is likely to be a fairly reliable indicator of any oscillations in this incumbent strength. It is not inconceivable that the party will be the first to show any serious schism or weakness in the authority of the current ruling group and so serve as a valuable early warning of regime breakdown.

The second related point to note is that United Russia, as a party reliant on the federal executive branch as a sponsor institution, is likely to find institutionalisation difficult, if not impossible. Comparative experience shows that parties reliant on other organisations/institutions not only take longer to acquire value, but also are prone to sudden collapse (Hopkin 1999; Panebianco 1988; Randall and Svasand 2002). Panebianco defines institutionalisation as the process by which an organisation 'incorporates its founders' values and aims', when the organisation makes a transition from a consumable entity with an instrumental nature to an institution. The organisation becomes 'valuable in and of itself, and its goals become inseparable and indistinguishable from it' – its 'preservation and survival become a "goal" for a great number of its supporters' (Panebianco 1988: 53). Ultimately, the fate of United Russia is tied to the fate of the federal executive branch, in particular to the figure of Vladimir Putin. The ability of United Russia to become a longer-term feature of Russian politics is still heavily reliant on the elite variable, as was the fate of every previous party of power.

At the same time, because of the impossibility of fully understanding the elite variable in Russia, it is possible that United Russia is currently being groomed by power-holders as a longer-term option for the regime and that at some point in the near future Russia's rulers will make an institutional transition and vest their power within the party. The quote at the beginning of this chapter sums up the longer-term hopes of many United Russia leaders and functionaries that the party will one day emerge as a real party of power. Former United Russia deputy, Viktor Grishin, sums up the party situation:

> We have created a party structure that can fight for power at all levels. I mean in both the representative and executive branches of power. But I cannot tell you that today we occupy all these levels of power. It isn't right to say this.
>
> (Grishin 2007)

The chair of United Russia's Central Executive Committee, Andrei Vorob'ev, echoes this belief, that United Russia will eventually become a genuine, principal power in the political system:

> Power is a matter of time. I have no doubt that very soon we will have president party-members and government party-members. This is just the matter of time. We must be ready, we must be organised and we must be sure in ourselves.
>
> (Vorob'ev 2007)

If the party were to gradually develop into a principal power in the political system, then this would not be without comparative precedent. As mentioned in Chapter 1, principal–agent relations are by no means static and appear to change over time. The Mexican PRI was created as a party very much dependent on power-holders in the aftermath of the Mexican Revolution, in a similar way to United Russia in the post-Yeltsin period. In the Mexican case it was a military and not a civilian executive that created the party, but nonetheless it took decades for the party to institutionalise and for the PRI to become more than just an agent of powerful generals in their attempt to control a diverse territory (Scott 1959).

However, to date, there has been little discernible movement towards this end in Russia, although, as mentioned at several junctures, the post-Yeltsin period has seen a gradual increase in the power of parties in general. Unfortunately, parties of power also provide useful 'buffers' for the federal executive branch if political circumstances spin out of control and this may be one reason why power-holders may choose to continue with the status quo. Even if United Russia were to gain more power in the political system, there is always a danger of unintended consequences, of the party becoming too confident and too independent. As far back as 2006, Kagarlitsky suggested that there were signs of this happening, noting that, 'having tasted victory, they are beginning to feel like real politicians and have forgotten that they are really nothing more than cardboard characters' (Kagarlitsky 2006). This may also make power-holders in and around the federal executive branch cautious in their dealings with the party of power in the future.

Ultimately, it remains to be seen if the post-Yeltsin political order is strong enough to perpetuate itself without the need of a ruling party of a classic, twentieth-century kind, seen in places such as Mexico, Malaysia, Taiwan and elsewhere. It also remains to be seen if United Russia will truly institutionalise to become a long-term feature of post-Soviet Russian politics. In the final analysis, it is worth considering the opinion of an insider with close connections with the federal executive branch and United Russia. On the subject of the prospects of Russia's rulers institutionalising their rule within the party of power, Sergei Markov comments: 'Yes, the President has thought about this many times, but every time he makes the decision – no, there is too much to lose, we will lose too much if we are members of the party' (Markov 2007). This quote seems particularly appropriate for understanding United Russia, its place in the post-Yeltsin period and the obstacles it faces in becoming a genuine ruling party.

Notes

Introduction

1 Geddes (2003: 78), using a dataset of authoritarian regimes (1946–2000), excluding those maintained by foreign occupation or military threat, found that the average length of rule in military regimes before a democratic transition was 9.5 years (33 cases), in personalist regimes 15.5 years (46 cases) and in single-party regimes 29 years (21 cases). Geddes also calculates the average length of rule for hybrid regimes combining elements of military, personalist and single-party rule, showing that the presence of a ruling party also contributes to regime longevity. For example, the average length of rule in military/personalist regimes without a party was only 11.3 years (12 cases), but for military/single-party and personalist/single-party regimes the average lifespan was 19.6 years (14 cases) and for triple hybrids 33 years (3 cases).
2 'Post-Soviet' is used to refer to all former Soviet Republics, with the exception of the three Baltic States that were incorporated into the Soviet Union during the Second World War. Post-communist refers to these Baltic States and all other former communist countries that were not part of the Soviet Union, but were at some point under its orbit of influence or shared similar political and economic systems.

1 Ruling parties and dominant-power politics

1 Huntington describes a 'democratic wave' as a group of transitions from non-democratic to democratic regimes that occur within a specified period of time and which significantly outnumber transitions in the opposite direction (Huntington 1993: 15). Although the circumstances of the third wave are open to debate (Diamond 1996), the current dynamic does suggest that this latest wave of democratisation has already finished. Between 1989 and 1999 the number of electoral democracies rose from 69 to 120, while from 2000 to 2009 the number dropped from 120 to 116. 1997 was the first year since the collapse of the Soviet Union when the number of electoral democracies in the world actually fell (Freedom House 2010). *The Economist*'s Democracy Index 2010 identifies 2008 as the start of a reverse wave (*The Economist* 2010: 1).
2 Chambers identifies several functions that US parties performed to help overcome a range of political 'loads' that appeared in the initial stages of democratisation, such as ensuring the effective operation of the national constitutional authority, expressing and aggregating conflicting interests in the new American society, as well as training and enlisting leaders and cadres (Chambers 1966: 90).
3 For details on Africa's transitions, see Bratton and van de Walle (1997: 79) and their overview of the five modal regimes that dominated postcolonial African politics.
4 Freedom House scores for political rights and civil liberties are based on surveys derived from the Universal Declaration of Human Rights and include 10 questions for

political rights and 15 questions for civil liberties. The ranking is based on a scale of 1–7. A score of 1 for political rights indicates that opposition plays an important role in that political system and has 'actual power', while a score of 7 indicates 'warlordism', violence and the absence of a central government. For civil liberties, states with a score of 1 are distinguished by an established and equitable system of rule of law, while states with a score of 7 have virtually no freedoms and an overwhelming and justified fear of repression. For details see Freedom House (2008).

5 At this juncture it is important to indicate how the terms 'patron–client relations', 'clientelism', 'patronage' and 'patrimonialism' are understood and used in the following chapters. Mainwaring (1999: 177) makes a distinction between patronage, clientelism and patrimonialism. Accordingly, patrimonialism is the use or distribution of state resources on a non-meritocratic basis for political gain. Patronage involves the distribution of jobs, work projects, state contracts and concessions, and state investments. Clientelism or patron–client relations is characterised by an unequal character, uneven reciprocity, a non-institutionalised nature and a face-to-face character. For the purpose of this research, the term 'patron–client relations' is used to cover all modes of exchange between electoral constituencies and politicians (Kitschelt and Wilkinson 2007: 7).

6 Magaloni does comment in a footnote that many hegemonic party autocracies are less prone to elite schisms because they do not replace their presidents as often (Magaloni 2006: 17 fn16).

7 Geddes (2003: 225–32) provides a coding system for classifying regimes as single party, military and personalist together with movements between regime types. Accordingly, Guinea, Guinea-Bissau and Mali are some of the examples of single-party rule becoming personalist, while Turkey 1923–83 shows a move from single-party to militarily rule.

8 Brooker notes that each twentieth-century dictatorship was established by an organisation, either a party or the military. As a result, the regime's leadership, whether an individual or political committee, acted as the agent of the ruling organisation, whether party or military. The author notes that, often, party or military rule was transformed into personal rule, ending and even reversing this principal–agent relationship (Brooker 1995: 10–11).

2 Parties of power

1 Alternative definitions of party of power are 'electoral blocs organised by state actors to participate in parliamentary elections and forge national organisations for presidential elections' (Smyth 2002: 556); 'parties which gain their position from support for the president, turning on its head the normative notion of the president gaining his position through the support of a party' (Bacon 2004: 42); 'party or quasi-party organisations created by elites to participate in elections' (Golosov and Likhtenshtein 2001: 7); 'something politically attuned Russians think of the way westerners think of the establishment' (Colton and McFaul 2003: 48); 'established by the executive branch in order to get a majority in legislative arenas; they lack any definite ideology; they shamelessly use state resources for campaigning and are merely captured by the top state officials' (Gel'man 2005: 5).

2 Russia's Choice had an impressive contingent of federal and regional elites including Egor Gaidar, Anatolii Chubais, Dmitrii Volkogonov and Sergei Filatov, among others. The PRES party-list included Sergei Shakhrai (Deputy Prime Minister and Committee on Nationalities), Aleksandr Shokhin (Deputy Representative of Government), Yurii Kalmykov (Minister of Justice) and Gennadii Melik'yan (Minster of Labour) (CEC 1993b).

3 Most definitions of 'parties of power' are single sentence. Oversloot and Verheul are perhaps the exception, as, unlike most commentators, they actually expand the party of

power definition to include a number of sub-categories, such as 'party of power helper parties', breaking this down further into 'satellite parties' and 'alternative parties of power' (Oversloot and Verheul 2006: 392). However, this strategy does not really help identify what a party of power is and is not.

4 The Russia's Choice bloc consisted mainly of the Democratic Russia Movement and the Russia's Choice Movement, but also several smaller organisations, including the Association of Private and Privatised Enterprises, the Peasant Party of Russia, the League of Co-operatives and Entrepreneurs, the New Party of Democratic Initiative, the Organisation of Peasants and Farmers of Russia, 'Living Circle' and the movement 'Military for Democracy' (McFaul 1998: 118).

5 One alternative option may be to change the Russian wording of the term to insert a preposition to form '*partiya 'u/pri' vlasti*' (party 'close to' power) or even '*partiya dlya vlasti*' (party 'for' power – see Vasil'tsov 2007). However, this strategy does not provide much more analytical clarity, although it is an improvement on 'party of power'.

6 The process of de-partification broadly began with the process of *perestroika* in 1986. However, Gorbachev sought to reduce the influence of the CPSU party elite, rather than the party per se (Harris 1997: 208). De-partification found expression in the 1993 constitution that weakened the formal role of parties and parliament. In post-Soviet Russia, de-partification, like in many countries of Central and Eastern Europe, continues to be relevant, seen in the way that parties are among the most distrusted institutions (Rose *et al.* 1998: 155; see also Chapter 3). Likewise, it has been noted that many candidates in elections in post-Soviet Russia in the 1990s would intentionally hide ties to parties to avoid alienating potential supporters (Moser 1999: 148).

7 For the circumstances surrounding the adoption of the constitution and the emergence of semi-presidentialism in Russia, see White (2000: 70–106).

8 Colton and Skach make the interesting observation that the Russian president has almost twice the constitutional strength of the president in the French Fifth Republic and at least a third more power than the president had in Weimar Germany (Colton and Skach 2005: 119).

9 This was exactly what happened in Russia. Smith calculated that President Yeltsin issued an average of 12–13 decrees per month before his clash with parliament in the autumn of 1993. This figure then increased to 65 decrees per month in December 1993, and by 1996, following the presidential election in the same year, Yeltsin had resorted to a de facto rule by decree (Smith 1996, cited in Lynch 2005: 142).

10 One of the earliest recorded references to the party of power comes from commentary on the Ukrainian parliamentary election of December 1993, where the term is used to describe the spectrum of parties supporting central power (Bogdanovskii 1993).

11 Brown notes that the elections to the CPD in 1989 were genuinely contested and ultimately undermined key pillars of communist rule, such as democratic centralism and the *nomenklatura* system of party appointments, while simultaneously introducing a considerable element of democracy and pluralism (Brown 1996: 181).

12 For example, in the course of the 1990s, the Russian national economy contracted by about a half, affecting the living standards of ordinary Russians and the state infrastructure they relied on (Lynch 2005: 3). In terms of the bombing of Serbia by NATO forces in 1999, a number of elites and experts interviewed highlighted this event as the starting point for increasing national sentiment and hostility towards the West, a vein of popular sentiment that both the post-Yeltsin administration and United Russia successfully tapped. One set of opinion poll data showed that over 60 per cent of respondents thought that NATO was responsible for the conflict in Yugoslavia in 1999 (Petrova 1999).

13 Polls conducted in November 1993, just prior to the December election, showed that around 30 per cent of those surveyed thought that the economic situation in Russia had become considerably worse since June of that year. However, only seven months

previously, Boris Yeltsin collected just under 59 per cent of the popular vote in the April 1993 referendum. It is therefore unclear why the results of shock therapy should influence the 1993 Duma election, but apparently not the April referendum (McFaul 1998: 123–4).

14 By post-Soviet Russian standards, the outcome of the 1996 presidential election was close run. In the first round, Zyuganov polled 32.5 per cent of the vote against Yeltsin's 35.8 per cent. In the second round, the result was more emphatic, with Yeltsin picking up 54.4 per cent and Zyuganov 40.7 per cent, although this belies the palpable sense of uncertainty surrounding these elections (Russia Votes, n.d.).

15 An interviewee who contributed on the condition of anonymity claimed to have been approached by Unity representatives in a hospital where she worked during the 1999 election campaign with the promise of cash in return for her vote. In general, the subject of electoral fraud in the 1999 State Duma election is difficult to gauge. One study suggests that the degree of falsification depended on governor support for either Unity or FAR (Ovchinnikov 2001).

3 United Russia and dominant-power politics

1 According to the Federal Service for State Statistics (Rosstat 2009), Russia's population as of January 2009 was approximately 141 million, with just over 50 million concentrated in 10 of the 83 subjects – Moscow City, St Petersburg, Chelyabinskaya Oblast', Krasnodarskii Krai, Moskovskaya Oblast', Rostovskaya Oblast', Sverdlovskaya Oblast', Tyumenskaya Oblast' and the national republics of Bashkortostan and Tatarstan.

2 Yanbukhtin makes the logical point that, of the 22 parties competing in the 2003 State Duma election, only a handful were serious competitors for United Russia in any one region. After identifying the nature and threat of these serious competitors in each of the regions, United Russia was then able to concentrate resources effectively (Yanbukhtin 2008: 44).

3 This information was provided in an internal party publication given to the author, containing instructions on how to use the party *konkursy* or competitions to best effect, as well as other media events to raise the profile of the party.

4 In the UK, for example, party membership stood at 9.4 per cent in 1964, but with the onset of antiparty sentiment it dropped to 2 per cent by 1992 (Webb 1996: 372).

5 According to a senior United Russia figure, the previous mandatory system saw party members of working age paying subscription fees at their local office in accordance with their income (Silantiev 2007). In general, up to 2008, indigent party members paid six roubles a quarter, while more wealthy members earning more than 20,000 roubles a month contributed 0.5 per cent of their salary in party fees (Zubchenko 2008).

6 This information was collated from the biographies and occupations of Duma deputies in the Fourth Duma Convocation (Segedinenko 2004).

7 This information was obtained from a personal invitation to a round table held in March 2007 in the editor's office of the Izvestiya Publishing House, Moscow. Participants included Sergei Markov, Aleksei Mitrofanov, Ivan Mel'nikov, Valerii Khomyakov, Vladimir Mamontov and Valerii Federov. The round table discussed a number of issues, including the electoral prospects of a number of political parties following regional elections in March and ahead of the December 2007 State Duma election.

8 Interestingly, Oates notes that 28 per cent of survey respondents said that the most important role of the media in Russia was to strengthen 'feelings of national unity among Russians' (Oates 2006: 150). This attitude may normalise biased television coverage of United Russia in the eyes of many voters and complement the unifying message of the party.

4 Governance in the post-Yeltsin period

1 A two-thirds majority or at least 300 deputies had to vote for any or all of the five charges levelled against the president for impeachment to commence. Results from the Duma voting on 15 May 1999 showed that opposition forces failed to muster 300 or more votes on any of the five points (Duma 1999). The third point of impeachment – war in Chechnya – was a main focus for the Yeltsin opposition (Tret'yakov 1999), but fell short with 284 votes.

2 The ODIHR has voiced its concerns over the practice of ahead of schedule voting at every election it has monitored in Belarus, 2001–10, despite changes made by the Belarus authorities to prevent its fraudulent use. In the 2001 presidential election, the ODIHR referred to the possibility that a mass substitution of votes had taken place in favour of Alexsandr Lukashenko as a result of the lax controls surrounding this early voting procedure (ODIHR 2001: 21). During the 2004 parliamentary election, the ODIHR criticised ahead of schedule voting for failing to provide conditions for voting secretly, because of an absence of polling booths. Observers also found individuals in possession of several ballot papers in the vicinity of polling stations (ODIHR 2004: 17). During early voting in the 2006 presidential election, there were reports that managers of companies and educational institutions had instructed their employees to vote early and, in some cases, threatened the cancellation of contracts or other penalties if they failed to do so (ODIHR 2006: 21).

3 In November 2009, during his address to the Federal Assembly, President Medvedev stated the need to take measures to prevent illegal manipulation during elections, with voting ahead of schedule singled out as one area in need of change (Medvedev 2009). As a result, the president forwarded a draft bill to the State Duma in January 2010, which was subsequently accepted by both chambers and signed into effect by Medvedev in late May 2010. Although this amendment restricts the scope of voting ahead of schedule, it remains to be seen how it will effect federal-level elections. The experience of Belarus suggests that stricter guidelines may not prevent fraud, while opponents of the amendment in Russia, among them United Russia Higher Council chair, Boris Gryzlov, claim that it will simply result in lower voter turnout (Edinaya Rossiya 2010a).

4 In January 2009, the Supreme Court published its decision on the dual complaints submitted by the Union of Right Forces in relation to the validity of the December 2007 State Duma election and the illegal seizure of its election campaign material. The court found that around 18 million items of campaign material were, in fact, seized illegally, but that this was insufficient grounds to reconsider the results of the election (Sova 2009b).

5 This calculation was made by the author based on the balance of power in the Fifth Duma Convocation using the formula proposed by Laakso and Taagepera (1979: 24).

6 According to the party website (Edinaya Rossiya 2010e), by September 2010, members of United Russia's Higher Council represented in presidential administration/government positions include Aleksandr Beglov (Deputy Chief of Staff in the Presidential Administration); Viktor Ishaev (Presidential Plenipotentiary Envoy); Oleg Goverun (Presidential Domestic Policy Directorate); Aleksandr Khloponin (Deputy Prime Minister); Sergei Sobyanin (Deputy Prime Minister); Aleksandr Zhukov (Deputy Prime Minister); Sergei Shoigu (Minister for Civil Defence); Elena Skrynnik (Minister for Agriculture); and Yurii Trutnev (Minister for Natural Resources and the Environment).

7 As of August 2010, the remaining six committees are chaired by A Just Russia (Science and Technology; Families, Women and Children), the LDPR (Cooperation between Independent States and Relations with Fellow Countrymen; Youth Matters) and the CPRF (Industry; Nationalities).

8 As of September 2010, the following members of United Russia's Higher Council sit on the Council for the Realisation of the National Projects: Vasilii Bochkarev, Georgii

Boos, Vladimir Chub, Aleksei Gordeev, Boris Gryzlov, Viktor Ishaev, Aleksandr Khloponin, Yurii Luzhkov, Valentina Matvienko, Oleg Morozov, Viktor Sadovnichii, Martin Shakkum, Aleksandr Shokhin, Sergei Sobyanin, Petr Sumin, Vyacheslav Volodin and Aleksei Zhukov. For the composition of the National Projects Council, see Rost (2010). For the composition of United Russia's Higher Council, see Edinaya Rossiya (2010e).

5 The politics of stability

1 To put this point in even more of a comparative perspective, of the 25 states comprising Central, Eastern and Southern Europe and of the 28 states comprising Asia – a total of 53 states – only six exist as federations (Derbyshire and Derbyshire 1999: 136–7, 328–9).
2 As of 2010, United Russia's website documenting its own party history from 2001 contains no reference to Bespalov, signifying the sensitivity of this subject.
3 According to a former Molodaya Gvardiya leader, Nadezhda Orlova (2007), party activists are entitled to compensation if they take part in certain activities, such as street protests.
4 The author struck up conversation with a young adult delivering United Russia election leaflets in a district of Moscow in the spring of 2007. This individual claimed that he was not a United Russia member and did not want to distribute United Russia's leaflets, but that his employer in a local mobile telephone company had instructed him to put on a United Russia jacket and cap and to deliver them anyway.
5 Figures for the Fourth State Duma Convocation were calculated using the biographical data from Segedinenko (2004). For the Fifth State Duma Convocation, biographical data from the official State Duma website were used.
6 This information was obtained from the official websites of the regions in question.

References

Abdullaev, N. and S. Saradzhyan (2006) 'Russia's Responses in the War on Terror: Legal, Public Policy, Institutional and Operational Strategies', in R. Orttung and A. Makarychev (eds), *National Counter-Terrorism Strategies: Legal, Institutional and Public-Policy Dimensions in the US, UK, France, Turkey and Russia*, Amsterdam: IOS Press, pp. 191–203.

Abente-Brun, D. (2009) 'Paraguay: The Unravelling of One-Party Rule', *Journal of Democracy*, 20 (1): 143–56.

Albrecht, H. (2007) 'Authoritarian Opposition and the Politics of Challenge in Egypt', in O. Schlumberger (ed.), *Debating Arab Authoritarianism*, Stanford: Stanford University Press, pp. 59–74.

Aldrich, J. (1995) *Why Parties? The Origins and Transformation of Political Parties in America*, Chicago: Chicago University Press.

Aleksandrov, V., O. Gorbunova and V. Troyanovskii (2005) 'Prodolzhayutsya Mitingi Protesta Protiv Zakona N 122', *Gazeta*. Online. Available at http://dlib.eastview.com/sources/article.jsp?id=7314276 (accessed 1 April 2009).

Alexander, M. (2004, February) 'Democratization and Hybrid Regimes: Comparative Evidence from Southeast Europe', paper presented at the Nuffield College Seminar on Democratization, Oxford.

Alksnis, V. (2007) Independent deputy, State Duma, interview with author, 13 February, Moscow.

Andrusenko, L. (1999) 'Berezovskii Ne Vidit Al'ternativy Putinu', *Nezavisimaya gazeta*. Online. Available at http://www.ng.ru/politics/1999–12–23/3_berezovsky.html (accessed 1 December 2008).

Angrist, M. (2006) *Party Building in the Modern Middle East*, Seattle: University of Washington Press.

Aron, L. (2007) *Russia's Revolution: 1989–2006*, Washington: AEI Press.

Auzan, A. (2007) Professor of Applied Institutional Economics, Moscow State University, interview with author, 23 July, Moscow.

Auzan, A. and P. Kryuchkovoi (2002) *Administrativnye Bar'ery v Ekonomike: Institutsional'nyi Analiz*, Moscow: IIF 'Spros' Konfop.

Bacon, E. (1998) 'Russia: Party Formation and the Legacy of the One-Party State', in J. White and P. Davies (eds), *Political Parties and the Collapse of Old Orders*, Albany: State University of New York Press, pp. 205–22.

Bacon, E. (2004) 'Russia's Law on Political Parties: Democracy by Decree', in C. Ross (ed.), *Russian Politics under Putin*, Manchester: Manchester University Press, pp. 39–52.

Bacon, E., B. Renz and J. Cooper (2006) *Securitising Russia: The Domestic Policies of Putin*, Manchester: Manchester University Press.

Balzer, H. (2003) 'Managed Pluralism: Vladimir Putin's Emerging Regime', *Post-Soviet Affairs*, 19 (3): 189–227.

Barabashev, A. (2008) 'United Russia and Bureaucratic Reform', e-mail, 12 December.

Barakhova, A. (2000) ' "Edinstvo" Ostalos' bez Shoigu', *Kommersant*. Online. Available at http://www.kommersant.ru/doc.aspx?DocsID=149107 (accessed 25 March 2011).

Belin, L. and R. Orttung (1997) *The Russian Parliamentary Elections of 1995: The Battle for the Duma*, London: M. E. Sharpe.

Belykh, N. (2007) Union of Right Forces, party leader, interview with author, 27 June, Moscow.

Bennett, A. and A. George (2001) 'Case Studies and Process Tracing in History and Political Science: Similar Strokes for Different Foci', in C. Elman and M. Elman (eds.), *Bridges and Boundaries: Historians, Political Scientists and the Study of International Relations*, London: MIT Press, pp. 137–66.

Bhaskar, R. (1978) *A Realist Theory of Science*, Brighton: Harvester.

Bielasiak, J. (1997) 'Substance and Process in the Development of Party Systems in East Central Europe', *Communist and Post-Communist Studies*, 30 (1): 23–44.

Bilevskaya, E. (2009) 'Konservatsiya Modernizatsii', *Nezavisimaya gazeta*. Online. Available at http://www.ng.ru/politics/2009-11-18/1_conservation.html (accessed 28 August 2011).

Bilevskaya, E. and I. Rodin (2008) 'Konstitutsiyu Popravyat k Ee Yubileyu', *Nezavisimaya gazeta*. Online. Available at http://www.ng.ru/politics/2008–11–26/1_ constitution.html (accessed 25 March 2011).

Blakkisrud, H. (2011) 'Medvedev's New Governors', *Europe-Asia Studies*, 63 (3): 369–97.

Blondel, J. (2000) 'Introduction', in J. Blondel and M. Cotta (eds), *The Nature of Party Government*, New York: Palgrave, pp. 1–17.

Bogdanovskii, V. (1993) 'Ukrainskaya Predvybornaya Mozaika: Risunok Ni Poluchaetsya', *Krasnaya zvezda*. Online. Available at http://dlib.eastview.com/sources/article.jsp?id=3363894 (accessed 5 January 2009).

Bolotova, O. (2010) 'Oboz Doekhal', *Gazeta*. Online. Available at http://www.gazeta.ru/politics/2010/04/26_a_3358001.shtml (accessed 24 March 2011).

Bondar', V. (2007) Director of the Institute for Local Self-Governance, interview with author, 31 May, Moscow.

Borisov, S. (1999) 'Vneshnepoliticheskaya Deyatel'nost' Rossiiskikh: Regionov Kak Atribut Ikh Politicheskoi Samoidentifikatsii', Moscow Carnegie Centre. Online. Available at http://www.carnegie.ru/ru/pubs/books/volume/56519.htm (accessed 4 June 2004).

Bowen, R. (2003) *Japan's Dysfunctional Democracy*, Armonk: M. E. Sharpe.

Bratton, M. and N. van de Walle (1997) *Democratic Experiments in Africa*, Cambridge: Cambridge University Press.

Brinegar, A., S. Morgenstern and D. Nielson (2006) 'The PRI's Choice: Balancing Democratic Reform and Its Own Salvation', *Party Politics*, 12 (1): 77–97.

Brooker, P. (1995) *Twentieth-Century Dictatorships*, New York: New York University Press.

Brooker, P. (2000) *Non-Democratic Regimes*, Basingstoke: Macmillan Press.

Brown, A. (1996) *The Gorbachev Factor*, New York: Oxford University Press.

Brownlee, J. (2007) *Authoritarianism in an Age of Democratization*, New York: Cambridge University Press.

Brudny, Y. (1998) 'St. Petersburg: The Election in the Democratic Metropolis', in T. Colton and J. Hough (eds), *Growing Pains: Russian Democracy and the Election of 1993*, Washington, D.C: Brookings Institution Press, pp. 349–96.

Brudny, Y. (2001) 'Continuity and Change in Russian Electoral Patterns? The December 1999 – March 2000 Electoral Cycle', in A. Brown (ed.), *Contemporary Russian Politics: A Reader*, Oxford: Oxford University Press, pp. 154–78.

Bunce, V. (1995) 'Should Transitologists Be Grounded?', *Slavic Review*, 54 (1): 111–27.

Bunce, V. (2003) 'Rethinking Recent Democratisation: Lessons from the Post-Communist Experience', *World Politics*, 55 (2): 167–92.

Bunce, V., M. McFaul and K. Stoner-Weiss (eds) (2009) *Democracy and Authoritarianism in the Post-Communist World*, New York: Cambridge Universtiy Press.

Bunich, A. (2007) Director of the international foundation 'Assistance for Business', interview with author, 12 October, Moscow.

Bunin, I. (2007) Director of the Centre of Political Technologies, interview with author, 23 March, Moscow.

Burnosov, Y. (2007) 'Pravitel'stvo i Partiya Ediny?', *Novaya Politika*. Online. Available at http://www.novopol.ru/text15005.html (accessed 25 March 2011).

Burton, M., R. Gunther and J. Higley (1992) 'Introduction: Elite Transformations and Democratic Regimes', in J. Higley and R. Gunther (eds), *Elites and Democratic Consolidation in Latin America and Southern Europe*, New York: Cambridge University Press, pp. 1–37.

Carothers, T. (2002) 'The End of the Transition Paradigm', *Journal of Democracy*, 13 (1): 5–21.

Casar, M. (2002) 'Executive–Legislative Relations: The Case of Mexico (1946–1947)', in S. Morgenstern and B. Nacif (eds), *Legislative Politics in Latin America*, Cambridge: Cambridge University Press, pp. 114–43.

Case, W. (1994) 'The UMNO Party Election in Malaysia: One for the Money', *Asian Survey*, 34 (10): 916–30.

Case, W. (1996a) 'Can the "Halfway House" Stand? Semidemocracy and Elite Theory in Three Southeast Asian Countries', *Comparative Politics*, 28 (4): 437–64.

Case, W. (1996b) 'UMNO Paramountcy: A Report on Single Party Dominance in Malaysia', *Party Politics*, 2 (1): 115–27.

CEC (1993a) 'Konstitutsiya Rossiiskoi Federatsii Ro', Tsentral'naya Izbiratel'naya Komissiya Rossiiskoi Federatsii. Online. Available at http://cikrf.ru/law/constitution.html (accessed 12 March 2011).

CEC (1993b) 'Vybory v Gosudarstvennuyu Dumu Federal'nogo Sobraniya Rossiisskoi Federatsii 1993 goda', Tsentral'naya Izbiratel'naya Komissiya Rossiiskoi Federatsii. Online. Available at http://cikrf.ru/banners/vib_arhiv/gosduma/1993/index.html (accessed 12 March 2011).

CEC (1995) 'Ob Obshchestvenykh Ob"edineniyakh', Tsentral'naya Izbiratel'naya Komissiya Rossiiskoi Federatsii. Online. Available at http://cikrf.ru/law/federal_law/zakon_82fed.html (accessed 12 March 2011).

CEC (2001) 'O Politicheskikh Partii', Tsentral'naya Izbiratel'naya Komissiya Rossiiskoi Federatsii. Online. Available at http://cikrf.ru/law/federal_law/2001_95fz.html (accessed 24 March 2011).

CEC (2002a) 'O Osnovnykh Garantiyakh Izbiratel'nykh Prav i Prava Na Uchastie v Referendume Grazhdan Rossiiskoi Federatsii', Tsentral'naya Izbiratel'naya Komissiya Rossiiskoi Federatsii. Online. Available at http://cikrf.ru/law/federal_law/zakon_02_67fz_n.html (accessed 24 March 2011).

CEC (2002b) 'O Protivodetsvii Ekstremistkoi Deyatel'nosti', Tsentral'naya Izbiratel'naya Komissiya Rossiiskoi Federatsii. Online. Available at http://cikrf.ru/law/federal_law/Zakon_02_114fz.html (accessed 24 March 2011).

CEC (2004) 'O Gosudarstvennoi Grazhdanskoi Sluzhbe Rossiiskoi', Tsentral'naya Izbiratel'naya Komissiya Rossiiskoi Federatsii. Online. Available at http://cikrf.ru/law/federal_law/zakon_79_fz.html (accessed 28 August 2011).

CEC (2005a) 'O Vyborakh Deputatov Gosudarstvennoi Dumy Federal'nogo Sobraniya Rossiiskoi Federatsii', Tsentral'naya Izbiratel'naya Komissiya Rossiiskoi Federatsii. Online. Available at http://cikrf.ru/law/federal_law/zakon_51.html (accessed 24 March 2011).

CEC (2005b) 'Ob Obshchestvennoi Palate Rossiiskoi Federatsii', Tsentral'naya Izbiratel'naya Komissiya Rossiiskoi Federatsii. Online. Available at http://cikrf.ru/law/federal_law/zakon_32_05fz.html (accessed 24 March 2011).

CEC (2008a) 'Postanovlenie', Tsentral'naya Izbiratel'naya Komissiya Rossiiskoi Federatsii. Online. Available at http://cikrf.ru/law/decree_of_cec/2008/01/22/Zp080693.html (accessed 24 March 2011).

CEC (2008b) 'Perechen' Zaregistrirovannykh Partii', Tsentral'naya Izbiratel'naya Komissiya Rossiiskoi Federatsii. Online. Available at http://www.cikrf.ru/politparty/ (accessed 24 March 2011).

CEC (2009a) 'O Resul'tatakh Proverki Svodnykh Finansovykh Otchetov Politicheskikh Partii o Postuplenii i Raskhodovanii Sredstv', Tsentral'naya Izbiratel'naya Komissiya Rossiiskoi Federatsii. Online. Available at http://www.cikrf.ru/politparty/finance/svodn_otchet.html (accessed 23 March 2011).

CEC (2009b) 'O Garantiyakh Ravenstva Parlamentskikh Partii pri Osveshchenii ikh Deyatel'nosti Gosudarstvennymi Obshchedostupnymi Telekanalami i Radiokanalami', Tsentral'naya Izbiratel'naya Komissiya Rossiiskoi Federatsii. Online. Available at http://www.cikrf.ru/law/federal_law/zakon_95fz_290609.html (accessed 23 March 2011).

CEC (2010a) 'O Vyborakh Prezidenta Rossiiskoi Federatsii', Tsentral'naya Izbiratel'naya Komissiya Rossiiskoi Federatsii. Online. Available at http://www.cikrf.ru/law/federal_law/zakon_19.html (accessed 24 March 2011).

CEC (2010b) 'O Postuplenii i Raskhodovanii Sredstv Politicheskikh Partii v IV Kvartale 2010 goda', Tsentral'naya Izbiratel'naya Komissiya Rossiiskoi Federatsii. Online. Available at http://cikrf.ru/politparty/finance/2010/index.html (accessed 23 March 2011).

CEC (2010c) 'Edinyi Den' Golosovaniya v Sub"ektakh Rossiiskoi Federatsii – 14 Marta 2010 goda', Tsentral'naya Izbiratel'naya Komissiya Rossiiskoi Federatsii. Online. Available at http://www.cikrf.ru/banners/vib_arhiv/electday/vib_140310/ (accessed 23 March 2011).

CEJ (2007) 'Vybory Deputatov Gosudarstvennoi Dumy RF 2007'. Online. Available at http://www.memo98.cjes.ru/diagrams/2007/200711-national.pdf (accessed 23 March 2011).

Chadaev, A. (2007) Senator in the Public Chamber, interview with author, 10 February, Moscow.

Chaisty, P. (2005) 'Majority Control and Executive Dominance: Parliament–President Relations in Putin's Russia', in A. Pravda (ed.), *Leading Russia: Putin in Perspective*. Oxford: Oxford University Press, pp. 119–37.

Chaisty, P. (2008) 'The Legislative Effects of Presidential Partisan Powers in Post-Communist Russia', *Government and Opposition*, 43 (3): 424–53.

Chambers, W. (1966) 'Parties and Nation-Building in America', in M. Weiner and J. LaPalombara (eds), *Political Parties and Political Development*, Princeton: Princeton University Press, pp. 79–107.

Chernov, D. (2003) 'Vremya Sobytii: Proval Nedoveriya', *Vremya*. Online. Available at http://dlib.eastview.com/sources/article.jsp?id=5030908 (accessed 4 October 2008).

Chernyavskii, A. and A. Samarina (2011) 'Antipartiinyi Vypad Alekseya Kudrina', *Nezavisimaya gazeta*. Online. Available at http://www.ng.ru/politics/2011–02–21/1_kudrin.html (accessed 25 March 2011).

Chhibber, P. and K. Kollman (2004) *The Formation of National Party Systems: Federalism and Party Competition in Canada, Great Britain, India and the United States*, Princeton, NJ: Princeton University Press.

Chu, Y.-H. (2001) 'The Legacy of One-Party Hegemony in Taiwan', in L. Diamond and R. Gunther (eds), *Political Parties and Democracy*, Baltimore: Johns Hopkins University Press, pp. 266–98.

Chugaev, S. (1995) 'Fraktsiya V. Zhirinovskogo Stremitsya Stat' Glavnoi Oporoi Prezidenta', *Izvestiya*. Online. Available at http://dlib.eastview.com/sources/article.jsp?id =3181140 (accessed 5 December 2008).

CIA (2010) 'World Fact Book'. Online. Available at https://www.cia.gov/library/publications/ the-world-factbook/geos/xx.html (accessed 1 December 2010).

Collier, D. and R. Adcock (1999) 'Democracy and Dichotomies: A Pragmatic Approach to Choices About Concepts', *Annual Review of Political Science*, 2: 537–65.

Colton, T. (1995) 'Superpresidentialism and Russia's Backward State', *Post-Soviet Affairs*, 11 (2): 144–9.

Colton, T. (2005) 'Putin and the Attenuation of Russia Democracy', in A. Pravda (ed.), *Leading Russia: Putin in Perspective*, Oxford: Oxford University Press, pp. 104–17.

Colton, T. (2008) *Yeltsin: A Life*, New York: Basic Books.

Colton, T. J. and M. McFaul (2000) 'Reinventing Russia's Party of Power: Unity and the 1999 Duma Election', *Post-Soviet Affairs*, 16 (3): 201–24.

Colton, T. and M. McFaul (2003) *Popular Choice and Managed Democracy: The Russian Elections of 1999 and 2000*, Washington, D.C: Brookings Institution Press.

Colton T. and C. Skach (2005) 'A Fresh Look at Semi Presidentialism: The Russian Predicament', *Journal of Democracy*, 16 (3): 113–26.

Cooper, J. (2003) 'Taiwan: Democracy's Gone Awry?', *Journal of Contemporary China*, 12 (34): 145–62.

Corrales, J. (2002) *Presidents without Parties: The Politics of Economic Reform in Argentina and Venezuela in the 1990s*, University Park: Pennsylvania State University Press.

CPRF (2010) 'Khronometrazh Partiinogo Teleefira Dekabr' 2010 goda', Kommunisticheskaya Partiya Rossiiskoi Federatsii. Online. Available at http://kprf.ru/tv/86429.html (accessed 23 March 2011).

Creswell, J. and R. Maietta (2002) 'Qualitative Research', in D. Miller and N. Salkind (eds), *Handbook of Research Design and Social Measurement*, Thousand Oaks: Sage, pp. 143–97.

Crotty, W. (2006) 'Party Origins and Evolution in the United States', in R. Katz and W. Crotty (eds), *Handbook of Party Politics*, London: Sage, pp. 25–33.

Daalder, H. (1966) 'Parties, Elites, and Political Developments in Western Europe', in M. Weiner and J. LaPalombara (eds), *Political Parties and Political Development*, Princeton: Princeton University Press, pp. 43–77.

D'Anieri, P. (2007) *Understanding Ukrainian Politics*, Armonk: M. E. Sharpe.

Dave, B. (2007) 'Nations in Transit, Kazakhstan', Freedom House. Online. Available at http://www.unhcr.org/refworld/pdfid/4756ad587.pdf (accessed 26 March 2011).

Davies, J. (2001) 'Spies as Informants: Triangulation and the Interpretation of Elite Interview Data in the Study of the Intelligence and Security Services', *Politics*, 21 (1): 73–80.

Dawisha, K. and S. Deets (2006) 'Political Learning in Post-Communist Elections', *East European Politics and Societies*, 20 (4): 691–728.

Denisov, V. (2003) 'Vybory-2003. Na Start Vyzyvayutsya', *Krasnaya zvezda*, 4 September, p. 2.

Derbyshire, J. and I. Derbyshire (1999) *Political Systems of the World*, Oxford: Helicon.

Diamond, L. (1996) 'Is the Third Wave Over?', *Journal of Democracy*, 7 (3): 20–37.

Diaz-Cayeros, A., B. Magaloni and B. Weingast (2003) 'Tragic Brilliance: Equilibrium Hegemony and Democratization in Mexico', Working Paper, Stanford University. Online. Available at http://papers.ssrn.com/sol3/papers.cfm?abstract_id= 1153510 (accessed 24 March 2011).

Dolgov, A. (2003) 'United Russia Conquerors the Air', *The Moscow Times*. Online. Available at http://www.themoscowtimes.com/stories/2003/09/24/001.html (accessed 6 July 2007).

Dorofeev, V. (1997) 'Borisom Bol'she, Borisom Men'she', *Vlast'*, 40 (246): 12.

Downs, A. (1957) *An Economic Theory of Democracy*, New York: Harper and Row.

Duma (1999) 'Stenogramma Zasedaniya 15 Maya 1999', Apparat Gosudarstvennoi Dumy. Online. Available at http://wbase.duma.gov.ru/steno/nph-sdb.exe (accessed 23 March 2011).

Duma (2001) 'Stenogramma Zasedaniya 21 Iyunya 2001', Apparat Gosudarstvennoi Dumy. Online. Available at http://wbase.duma.gov.ru/steno/nph-sdb.exe (accessed 23 March 2011).

Duma (2005) 'Stenogramma Zasedaniya 22 Aprelya 2005', Apparat Gosudarstvennoi Dumy. Online. Available at http://wbase.duma.gov.ru/steno/nph-sdb.exe (23 March 2011).

Duma (2007) 'Stenogramma Zasedaniya 19 Fevralya 2007', Apparat Gosudarstvennoi Dumy. Online. Available at http://transcript.duma.gov.ru/node/757/?full (accessed 23 March 2011).

Duma (2010a) 'Zamestiteli Predsedatelya Gosudarstvennoi Dumy', Gosudarstvennaya Duma. Online. Available at http://www.duma.gov.ru/structure/leaders/vice-chairmans/ (accessed 25 March 2011).

Duma (2010b) 'Fraktsiya Vserossiiskoi Politicheskoi Partii "Edinaya Rossiya" ', Gosudarstvennaya Duma. Online. Available at http://www.duma.gov.ru/structure/factions/er/ (accessed 25 March 2011).

Duma (2011) 'Statisticheskie Dannye o Khode i Rezul'tatakh Zakonodatel'nogo Protsessa', Gosudarstvennaya Duma. Online. Available at http://www.duma.gov.ru/legislative/statistics/ (accessed 25 March 2011).

Duverger, M. (1964) *Political Parties: Their Organisation and Activities in the Modern State*, London: Methuen.

Edinaya Rossiya (2002) 'Programma Vserossiiskoi Politicheskoi Partii "Edinstvo" i "Otechestvo" Edinaya Rossiya'. Online. Available at http://www.edinros.ru/section.html?rid=48 (accessed 22 July 2004).

Edinaya Rossiya (2003a) 'Put' Natsional'nogo Uspekha: Manifest Partii Edinaya Rossiya'. Online. Available at www.edinros.ru/section.html?rid=145 (accessed 22 July 2004).

Edinaya Rossiya (2003b) 'Predvybornaya Programma Politicheskoi Partii Edinaya Rossiya'. Online. Available at www.edinros.ru/section.html?rid=2092 (accessed 22 July 2004).

Edinaya Rossiya (2006) 'V Ul'yanovskoi Oblasti Sformirovannoe Partiinoe Pravitel'stvo'. Online. Available at http://er-portal.ru/text.shtml?3/1309,101876 (accessed 22 March 2011).

Edinaya Rossiya (2007a) 'Partiya Edinaya Rossiya segodnya'. Online. Available at http://www.edinros.ru/er/rubr.shtml?110103 (accessed 21 January 2011).

Edinaya Rossiya (2007b) 'Regional'nye Otdeleniya'. Online. Available at http://www.er.ru/news.html?rid=3122 (accessed 1 September 2007).

Edinaya Rossiya (2007c) 'Plan Putina – Dostoinoe Budushchee Velikoi Strany'. Online. Available at http://www.er.ru/rubr.shtml?110099 (accessed 2 January 2011).

Edinaya Rossiya (2008a) 'Partiya Edinaya Rossiya Segodnya'. Online. Available at http://www.er.ru/rubr.shtml?110099 (accessed 5 July 2009).

Edinaya Rossiya (2008b) 'Edinaya Rossiya 2008 Obshchestvennye Ob"edineniya'. Online. Available at http://www.er.ru/text.shtml?3/6782 (accessed 23 March 2011).

Edinaya Rossiya (2008c) 'MKAPP dlya 'Edinoi Rossii' – Perspektivnyi i Konstruktivnyi Partnep'. Online. Available at http://www.edinros.ru/text.shtml?5/4072,100063 (accessed 23 March 2011).

Edinaya Rossiya (2009a) 'Sochi – 2014'. Online. Available at http://www.er.ru/rubr.shtml?110132 (accessed 23 March 2011).

Edinaya Rossiya (2009b) 'S"ezd Initsiiruet Novyi Partiinyi Proekt'. Online. Available at http://www.edinros.ru/news.html?id=148596 (accessed 7 January 2009).

Edinaya Rossiya (2009c) 'Rossiya: Sokhranim i Priumnozhim!'. Online. Available at http://www.edinros.ru/text.shtml?10/9535,110746 (accessed 23 March 2011).

Edinaya Rossiya (2009d) 'Chto Nuzhno dlya togo Chtoby Stat' Chlenom Partii Edinaya Rossiya'. Online. Available at http://edinros.ru/er/rubr.shtml?802 (accessed 23 March 2011).

Edinaya Rossiya (2009e) 'Ustav Vserossiiskoi Politicheskoi Partii Edinaya Rossiya'. Online. Available at http://edinros.ru/er/rubr.shtml?110102#1312 (accessed 23 March 2011).

Edinaya Rossiya (2009f) 'Stenogramma Prezentatzii Proekta Partii Edinaya Rossiya' 'Mirovoi Opyt Konservativnoi Modernizatsii'. Online. Available at http://www.er.ru/text. shtml?11/1107 (accessed 23 March 2011).

Edinaya Rossiya (2009g) 'Boris Gryzlov: Ideologiya Rossiiskogo Konservatizma – Eto Idiologiya Zaboty Obo Vsekh Grazhdanakh Strany'. Online. Available at http://edinros.ru/ er/text.shtml?10/9148,110031 (accessed 23 March 2011).

Edinaya Rossiya (2010a) 'Chleny Soveta Federatsii – Chleny Partii'. Online. Available at http:// edinros.er.ru/er/rubr.shtml?110098 (accessed 23 March 2011).

Edinaya Rossiya (2010b) 'Ustav Vserossiiskoi Politicheskoi Partii Edinaya Rossiya'. Online. Available at http://edinros.er.ru/er/rubr.shtml?110102#7 (accessed 23 March 2011).

Edinaya Rossiya (2010c) 'Ya Protiv Otmeny Dosrochnogo Golosovaniya'. Online. Available at http://er.ru/text.shtml?11/9152 (accessed 25 March 2011).

Edinaya Rossiya (2010d) 'Vyshi Sovet Vserossiiskoi Politicheskoi Partii Edinaya Rossiya'. Online. Available at http://edinros.er.ru/er/rubr.shtml?110090 (accessed 25 March 2011).

Edinaya Rossiya (2010e) 'Severo-Kavkazskii Mezhregional'nyi Koordinatsionnyi Sovet Partii Edinaya Rossiya'. Online. Available at http://edinros.er.ru/er/rubr.shtml?110887 (accessed 25 March 2011).

Edinaya Rossiya (2010f) 'Istoricheskaya Pamyat' '. Online. Available at http://www.edinros.ru/ rubr.shtml?110063 (accessed 25 March 2011).

Edinaya Rossiya (2010g) 'Istoriaya Partii Edinaya Rossiya'. Online. Available at http://www. er.ru/rubr.shtml?110112 (accessed 25 March 2011).

Edinaya Rossiya (2011a) 'Predsedateli i Glavnye Spetsialisty Regional'nykh Kontrol'no-Revizionnikh Komissii'. Online. Available at http://www.edinros.ru/text.shtml?6/6440 (accessed 23 March 2011).

Edinaya Rossiya (2011b) 'Adresa Regional'nykh Otdelenii Partii'. Online. Available at http:// www.er.ru/rubr.shtml?110107 (accessed 23 March 2011).

Edinstvo (1999) 'Tezisy k Izbiratel'noi Programme Bloka Medved'. Online. Available at http:// www.panorama.ru/works/vybory/party/p-med.html (accessed 25 March 2011).

Eisenhardt, K. (1989a) 'Building Theories from Case Study Research', *Academy of Management Review*, 14 (4): 532–50.

Eisenhardt, K. (1989b) 'Agency Theory: An assessment and Review', *Academy of Management Review*, 14 (1): 57–74.

Eldersveld, S. (1964) *Political Parties: A Behavioural Analysis*, Chicago: Rand McNally.

Elster, J. (1989) *Nuts and Bolts for the Social Sciences*, Cambridge: Cambridge University Press.

Emel'yanov, M. (2007) United Russia deputy, State Duma, interview with author, 26 June, Moscow.

Enyedi, Z. (2006) 'Party Politics in Post Communist Transition', in R. Katz and W. Crotty (eds), *Handbook of Party Politics*, London: Sage, pp. 228–38.

Epstein, L. (1980) *Political Parties in Western Democracies*, New Brunswick: Transaction.

Ermolin, V. (1993) 'Bloki, Partii, Dvizheniya: Kto i s Chem Vykhodit Na Start Predvybornoi Gonki?', *Krasnaya zvezda*. Online. Available at http://dlib.eastview.com/sources/article. jsp?id=3365623 (accessed 5 January 2009).

Evans, G. and S. Whitefield (1993) 'Identifying the Bases of Party Competition in Eastern Europe', *British Journal of Political Science*, 23 (4): 521–48.

Ezrow, N. and E. Frantz (2011) *Dictators and Dictatorships: Understanding Authoritarian Regimes and their Leaders*, New York: Continuum.

Fainsod, M. (1963) *How Russia Is Ruled*, Cambridge: Harvard University Press.

Falleti, T. and J. Lynch (2009) 'Context and Causal Mechanisms in Political Analysis', *Comparative Political Studies*, 42 (9): 1143–66.

Farizova, S. (2003) 'Anatoliya Chubaisa Spasla Partiya Vlasti', *Kommersant*. Online. Available at http://www.kommersant.ru/doc.aspx?DocsID=411062 (accessed 25 March 2011).

Farrell, D. (2002) 'Campaign Modernization and the West European Party', in K. Luther and F. Muller-Rommel (eds), *Political Parties in the New Europe: Political and Analytical Challenges*, Oxford: Oxford University Press, pp. 63–83.

Fedorov, V. (2007) *Putevoditel'po Vyboram*, Moscow: VTsIOM.

Felshtinsky, Y. and V. Pribylovskii (2008) *The Age of the Assassins: The Rise and Rise of Vladimir Putin*, London: Gibson Square.

Fish, S. (1995) *Democracy from Scratch: Opposition and Regime in the New Russian Revolution*, Princeton, NJ: Princeton University Press.

Fish, S. (2005) *Democracy Derailed in Russia*, Cambridge: Cambridge University Press.

Fleron, F. and R. Ahl (1998) 'Does the Public Matter for Democratization in Russia? What Have We Learned from "Third Wave" Transitions and Public Opinion Surveys?', in H. Eckstein, F. Fleron, E. Hoffman and W. Reisinger (eds), *Can Democracy Take Root In Post-Soviet Russia*, Oxford: Rowman & Littlefield, pp 287–96.

FNPR (2010) 'Chlenskie Organizatsii', Federatsiya Nezavisimykh Profsoyuzov Rossii. Online. Available at http://fnpr.org.ru/n/251/ (accessed 25 March 2011).

FOM (2000) 'In your View, Does the Present Situation in Chechnya Make Putin More Popular or Neither More Nor Less Popular?'. Online. Available at http://bd.english.fom.ru/report/cat/societas/regions/n_caucasus/putin_chechnya/etm000933 (accessed 10 January 2009).

FOM (2003a) 'Russian Voting Behaviour and Election Results'. Online. Available at http://bd.english.fom.ru/report/cat/policy/party_rating/stat_rating/party_ratings_2003/ed033401 (accessed 17 September 2008).

FOM (2003b) 'Ratings'. Online. Available at http://bd.english.fom.ru/report/cat/policy/party_rating/stat_rating/party_ratings_2003/ed033401 (accessed 17 September 2008).

FOM (2006) 'Attitudes towards the Party'. Online. Available at http://bd.english.fom.ru/report/cat/policy/edinaya_rossiya/etb064108 (accessed 5 April 2007).

FOM (2007a) 'Otnoshenie k Vyboram i Motivatsiya Neuchastiya v Nikh'. Online. Available at http://bd.fom.ru/report/cat/elect/est_el/d072722 (accessed 17 March 2011).

FOM (2007b) 'Ratings'. Online. Available at http://bd.english.fom.ru/report/cat/policy/party_rating/stat_rating/party_ratings_2003/ed033401 (accessed 17 September 2008).

FOM (2007c) 'Election Procedures in Russia'. Online. Available at http://bd.english.fom.ru/report/map/ed072123 (accessed 17 September 2008).

FOM (2008a) 'Edinaya Rossiya: Kto i Pochemu Vstupaet v Partiinye Ryady'. Online. Available at http://bd.fom.ru/report/cat/polit/pol_par/er/d082122 (accessed 18 March 2011).

FOM (2008b) 'Otnoshenie k Edinoi Rossii'. Online. Available at http://bd.fom.ru/report/cat/polit/pol_par/er/d081621 (accessed 18 March 2011).

FOM (2009) 'Deyatel'nost' Partii Edinaya Rossiya v Situatsii Ekonomicheskogo Krizisa'. Online. Available at http://bd.fom.ru/report/cat/polit/pol_par/er/d090612 (accessed 18 March 2011).

Fossato, F., J. Lloyd and A. Verhovsky (2008) *The Web That Failed: How Opposition Politics and Independent Initiatives are Failing on the Internet in Russia*, Oxford: Reuters Institute for the Study of Journalism.

Freedom House (2008) 'Methodology'. Online. Available at http://www.freedomhouse.org/template.cfm?page=351&ana_page=341&year=2008 (accessed 24 March 2011).

Freedom House (2010) 'Combined Average Ratings – Independent Countries'. Online. Available at http://www.freedomhouse.org/template.cfm?page=546&year= 2010 (accessed 24 March 2011).

Freedom House (2011) 'Freedom in the World Comparative and Historical Data'. Online. Available at http://www.freedomhouse.org/template.cfm?page=439 (accessed 24 March 2011).

Freeland, C. (2000) *Sale of the Century: The Inside Story of the Second Russian Revolution*, London: Little, Brown.

Frolov, A. (2003) *Biblioteka Edinoi Rossii 'Deistviya'*, Moscow: Algorithm.

Gaidar, E. (2007) Director of the Institute of the Economy in Transition, interview with author, 28 May, Moscow.

Gamov, A. (2004) 'Kudrin i Zhukov Vstanyut pod Znamena "Edinorossov" ', *Komsomol'skaya Pravda*. Online. Available at http://www.kp.ru/daily/23381/32959/ (accessed 25 March 2011).

Gandhi, J. and A. Przeworksi (2006) 'Cooperation, Cooptation, and Rebellion Under Dictatorships', *Economics & Politics*, 18 (1): 1–26.

Geddes, B. (1995) 'A Comparative Perspective on the Leninist Legacy in Eastern-Europe', *Comparative Political Studies*, 28 (2): 239–74.

Geddes, B. (1999) 'What Do We Know about Democratization after 20 Years?', *American Review of Political Science*, 2: 115–44.

Geddes, B. (2003) *Paradigms and Sand Castles: Theory Building and Research Design in Comparative Politics*, Ann Arbor: University of Michigan Press.

Geddes, B. (2006) 'Why Parties and Elections in Authoritarian Regimes', revised paper prepared for the annual meeting of the American Political Science Association, Washington, D.C., 2005. Available at http://www.daniellazar.com/wp-content/uploads/authoritarian-elections. doc (accessed 19 August 2011).

Gel'man, V. (2004) 'The Unrule of Law in the Making: The Politics of Informal Institution Building in Russia', *Europe-Asia Studies*, 56 (7): 1021–40.

Gel'man, V. (2005, October) 'From Feckless Pluralism to Dominant Power Politics? The Transformation of Russia's Party System', paper presented at the Ohio State University Conference on Post-Soviet In/securities: Theory and Practice, Columbus.

Gel'man, V. (2006) 'Perspektivy Dominiruyushchei Partii v Rossii', *Pro et Contra*, July/August: 62–71.

Gel'man, V. (2008) 'Party Politics in Russia: From Competition to Hierarchy', *Europe-Asia Studies*, 60 (9): 913–30.

Gel'man, V. and G. Golosov (1998) 'Regional Party Sytem Formation in Russia: The Deviant Case of Sverdlovsk Oblast', *Journal of Communist Studies and Transition Poltics*, 14 (1): 31–53.

George, A. and A. Bennett (2005) *Case Studies and Theory Development in the Social Sciences*, London: MIT Press.

George, C. (2007) 'Media in Malaysia: Zone of Contention', *Democratization*, 14 (5): 893–910.

Gevorkyan, N., N. Timakova and A. Kolesnikov (2000) *First Person: An Astonishingly Frank Self-Portrait by Russia's President*, London: Hutchinson.

Godzimirski, J. (2001) 'Fedrelandet-Hele Russland', Norwegian Institute of International Affairs. Online. Available at http://www2.nupi.no/cgi-win//Russland/polgrupp. exe?Fatherland-All+Russia (accessed 3 June 2008).

Goldman, S. (2002) 'Russian National Security after September 11', Report for Congress (RL31543). Online. Available at http://digital.library.unt.edu/govdocs/crs/permalink/meta-crs-7029 (accessed 26 March 2011).

Golosov, G. (1999a) 'The Origins of Contemporary Russian Political Parties: 1987-1993', in V. Gel'man and G. Golosov (eds), *Elections in Russia, 1993–1996: Analyses, Documents and Data*, Berlin: WZB, pp. 73–98.

Golosov, G. (1999b) 'Political Parties in the 1993–1996 Elections', in V. Gel'man and G. Golosov (eds), *Elections in Russia, 1993–1996: Analyses, Documents and Data*. Berlin: WZB, pp. 99–126.

Golosov, G. (2004) *Political Parties in the Regions of Russia: Democracy Unclaimed*, London: Lynne Rienner.

Golosov, G. (2008) 'Elektoral'nyi Avtoritarizm v Rossii', *Pro et Contra*, January/February: 22–35.

Golosov, G. and A. Likhtenshtein (2001) 'Partii Vlasti I Rossiiskii Institutsional'nyi Dizain: Teoriticheskii Analiz', *Polis*, 1: 6–14.

Gomzikova, S. and I. Rodin (2007) 'Podozritel'naya Kniga Nashla Chitatelei', *Nezavisimaya gazeta*. Online. Available at http://www.ng.ru/events/2007–10–31/12_book.html (accessed 23 March 2011).

Graney, K. (1997) 'Russia Bashkortostan: A Case Study on Building National Identity', Radio Free Europe/Radio Liberty (RFE/RL). Online. Available at http://www.rferl.org/content/article/1086129.html (accessed 25 March 2011).

Greene, K. (2007) *Why Dominant Parties Lose: Mexico's Democratization in Comparative Perspective*, New York: Cambridge University Press.

Grishin, V. (2007) United Russia deputy, State Duma, interview with author, 7 March, Moscow.

Gromov, A. (2007) 'Dikaya i Pouchitel'naya istoriya co Znamenem Pobedy', *Ekspert*. Online. Available at http://expert.ru/2007/04/12/pobeda (accessed 12 February 2011).

Gudkov, G. (2007) A Just Russia deputy, State Duma, interview with author, 21 February, Moscow.

Gudkov, L. and B. Dubin (2007) 'Posttotalitarnyi Sindrom: "Upravlyaemaya Demokratiya" i Apatiya Mass', in M. Lipman and A. Ryabov (eds), *Puti Rossiiskogo Postkommunizma*, Moscow: Carnegie Endowment for International Peace, pp. 8–63.

Gunther, R. and L. Diamond (2003) 'Species of Political Parties: A New Typology', *Party Politics*, 9 (2): 167–99.

Gurov, A. (2007) United Russia deputy, State Duma, interview with author, 19 April, Moscow.

Gurova, T. and Y. Polunin (2007) 'Reiting Politicheskoi Vliyatel'nosti', *Ekspert*, 12: 20–55.

Guy Peters, G. (2000) *Institutional Theory in Political Science*, London: Continuum.

Gvosdev, N. (2002) 'Mexico and Russia: Mirror Images?', *Demokratizatsiya*, 10 (4): 488–508.

Hadenius, A. and J. Teorell (2007) 'Pathways from Authoritarianism', *Journal of Democracy*, 18 (1): 143–56.

Hale, H. (2003) 'Explaining Machine Politics in Russia's Regions: Economy, Ethnicity and Legacy', *Post-Soviet Affairs*, 19 (3): 228–63.

Hale, H. (2006) *Why Not Parties in Russia? Democracy, Federalism and the State*, New York: Cambridge University Press.

Hale, H., M. McFaul and T. Colton (2004) 'Putin and the Delegative Democracy Trap: Evidence from Russia's 2003–2004 Elections', *Post- Soviet Affairs*, 20 (4): 285–319.

Hanson, S. (2001) 'Defining Democratic Consolidation', in R. Anderson, S. Fish, S. Hanson and P. Roeder (eds), *Post-Communism and the Theory of Democracy*, Princeton: Princeton University Press, pp. 126–51.

Harmel, R. and K. Janda (1994) 'An Integrated Theory of Party Goals and Party Change', *Journal of Theoretical Politics*, 6 (3): 259–87.

Harris, J. (1997) 'President and Parliament in the Russian Federation', in K. Von Mettenheim (ed.), *Presidential Institutions and Democratic Politics: Comparing Regional and National Contexts*, Baltimore: Johns Hopkins University Press, pp. 204–36.

Hay, C. (1996) *Re-Stating Social and Political Change*, Buckingham: Open University Press.

Hay, C. (2002) *Political Analysis*, Basingstoke: Palgrave.

Henkin, S. (1996) ' "Partiya Vlasti': Rossiiskii Variant', *Pro et Contra*, 1: 34–43.

Heryanto, A. (2003) 'Public Intellectuals, Media and Democratization: Cultural Politics of the Middle Classes in Indonesia', in A. Heryanto and S. Mandal (eds), *Challenging Authoritarianism in Southeast Asia*, New York: RoutledgeCurzon, pp. 24–59.

Heydemann, S. (1999) *Authoritarianism in Syria: Institutions and Social Conflict 1946–1970*, New York: Cornell University Press.

Hoffmann-Lange, U. (1987) 'Surveying National Elites in the Federal Republic of Germany', in G. Moyser and M. Wagstaffe (eds), *Research Methods for Elites Studies*, London: Allen & Unwin, pp. 27–47.

Holt, G., P. Love and H. Li (2000) 'The Learning Organisation: Toward a Paradigm for Mutually Beneficial Strategic Construction Alliances', *International Journal of Project Management*, 18: 415–21.

Hopkin, J. (1999) *Party Formation and Democratic Transition in Spain: The Creation and Collapse of the Union of the Democratic Centre*, Basingstoke: Macmillan.

Hopkin, J. (2001) 'A Southern Model of Electoral Mobilisation? Clientelism and Electoral Politics in Spain', *West European Politics*, 24 (1): 115–36.

Huat, C. (2007) 'Political Culturalism, Representation and the People's Action Party of Singapore', *Democratization*, 14 (5): 911–27.

Hughes, J. (2001) 'From Federalisation to Recentralisation', in S. White, A. Pravda and Z. Gitelman (eds), *Developments in Russian Politics 5*, Basingstoke: Pelgrave, pp. 128–46.

Huntington, S. (1968) *Political Order in Changing Societies*, New Haven, CT: Yale University Press.

Huntington, S. (1991–2) 'How Countries Democratize', *Political Science Quarterly*, 106 (4): 579–616.

Huntington, S. (1993) *The Third Wave: Democratization in the Late Twentieth Century*, Norman: University of Oklahoma Press.

Huskey, E. (2001) 'Democracy and Institutional Design in Russia', in A. Brown (ed.), *Contemporary Russian Politics: A Reader*, Oxford: Oxford University Press, pp. 29–45.

IIDEA (n.d.) Voter Turnout Data for Russian Federation'. Online. Available at http://www.idea.int/vt/country_view.cfm? CountryCode=RU (accessed 26 March 2011).

Il'nitskii, A. (2007) Head of Supporter Relations, United Russia Central Executive Committee, interview with author, 15 June, Moscow.

Ishiyama, J. and R. Kennedy (2001) 'Superpresidentialism and Political Party Development in Russia, Ukraine, Armenia and Kyrgyzstan', *Europe-Asia Studies*, 53 (8): 1177–91.

Ivanov, V. (2008) *Partiya Putina: Istoriya Edinoi Rossii*, Moscow: Ulma.

Jack, A. (2004) *Inside Putin's Russia: Can There Be Reform without Democracy?*, Oxford: Oxford University Press.

Jensen, D. (1999) 'How Russia Is Ruled – 1998', *Demokratizatsiya*, 7 (3): 341–69.

Jessop, B. (1990) *State Theory: Putting Capitalist States in their Place*, Cambridge: Polity Press.

Kagarlitsky, B. (2003) 'Red-White United Russia', *The Moscow Times*. Online. Available at http://www.themoscowtimes.com/stories/2003/12/04/009-print.html (accessed 16 September 2005).

Kagarlitsky, B. (2006) 'Power to the Party of Power', *The Moscow Times*, 23 March, p. 9.

Kalyvas, S. (1996) *The Rise of Christian Democracy in Europe*, New York: Cornell University Press.

Katz, R. (1997) *Democracy and Elections*, Oxford: Oxford University Press.

Katz, R. and P. Mair (1995) 'Changing Models of Party Organisation and Party Democracy: The Emergence of the Cartel Party', *Politics*, 1 (1): 5–28.

Kazantseva, S. (2008) 'Newspapers in Oblivion', Johnson's Russia List 2008-#120. Online. Available at http://www.cdi.org/russia/johnson (accessed 6 July 2008).

Key, V. (1955) 'A Theory of Critical Elections', *Journal of Politics*, 17 (1): 3–18.

Khamraev, V. (2010) ' "Edinaya Rossiya" Nashla Antipartinogo Liderai', *Kommersant*. Online. Available at http://www.kommersant.ru/Doc/1405805 (accessed March 23 2011).

Khamraev, V. and A. Barakhova (2008) 'Partiinoe Pravitel'stvo Obkatayut k 2012 gody', *Kommersant*. Online. Available at http://kommersant.ru/Doc/883007 (accessed 25 March 2011).

King, S. (2010) *The New Authoritarianism in the Middle East and Africa*, Bloomington: Indiana University Press.

Kiser, E. (1999) 'Comparing Varieties of Agency Theory in Economics, Political Science and Sociology: An Illustration from State Policy Implementation', *Sociological Theory*, 17 (2): 146–70.

Kitschelt, H. (1995) 'Formation of Party Cleavages in Post-Communist Democracies: Theoretical Propositions', *Party Politics*, 1 (4): 447–72.

Kitschelt, H. and S. Wilkinson (2007) 'Citizen–Politician Linkages: An Introduction', in H. Kitschelt and S. Wilkinson (eds), *Patrons, Clients, and Policies: Patterns of Democratic Accountability and Political Competition*. Cambridge: Cambridge University Press, pp. 1–49.

Kitschelt, H., Z. Mansfeldova, R. Markowski and G. Toka (1999) *Post-Communist Party Systems: Competition, Representation and Inter-Party Cooperation*, New York: Cambridge University Press.

Klimova, S. (2003) 'Political Parties and TV Advertising', FOM. Online. Available at http://bd.english.fom.ru/report/cat/societas/mass_media/election_feature/ed034723 (accessed 2 October 2008).

Klimova, S. (2006) 'United Russia in the Regions', FOM. Online. Available at http://bd.english.fom.ru/report/map/ed064122 (accessed 2 October 2008).

Kononenko, V. (1996) 'El'tsin i Chernomyrdin Mogut Vkluchit'sya v Predvybornuyu Kampaniyu Odnovremenno', *Izvestiya*. Online. Available at http://dlib.eastview.com/sources/article.jsp?id=3189926 (accessed 5 January 2009).

Kopecky, P. (1995) 'Developing Party Organizations in East-Central Europe: What Type of Party Is Likely to Emerge?', *Party Politics*, 1 (4): 515–34.

Korgunyuk, Yu. and S. Zaslavskii (1996) *Rossiiskaya Mnogopartiinost': Stanovlenie, Funktsionirovanie, Razvitie*, Moscow: INDEM.

Kostikov, V. (2007) 'Pochemu Putin Ne Stal Stalinym?', *Argumenti i Fakti Ukraina*. Online. Available at http://www.ukr.aif.ru/society/article/14331 (accessed 23 March 2011).

Kostyukov, A. (2003) '100 Vedushchikh Politikov Rossii v Mae', *Nezavisimaya gazeta*. Online. Available at http://dlib.eastview.com/sources/article.jsp?id=4980380 (accessed 5 December 2008).

Kotova, Yu. (2010) 'Skandal v Voronezhe: "Edinaya Rossiya" Boretsya s Pazharami s Pomoshch'yu Photoshop', MyWebs. Online. Available at http://mywebs.su/blog/life/1909.html (accessed 25 March 2011).

Krasner, S. (1984) 'Approaches to the State: Alternative Conceptions and Historical Dynamics', *Comparative Politics*, 16 (2): 223–46.

Krutov, A. (2007) United Russia deputy, Moscow Legislative Assembly, interview with author, 2 March, Moscow.

Kryshtanovskaya, O. and S. White (2005) 'Inside the Putin Court: A Research Note', *Europe-Asia Studies*, 57 (7): 1065–75.

Kubicek, P. (1994) 'Delegative Democracy in Russia and Ukraine', *Communist and Post-Communist Studies*, 27 (4): 423–41.

Kudinov, O. and G. Shipilov (1997) *Dialektika Vyborov*, Moscow: ZAO PO 'Master'.

Kudrin, A. (2005) 'Aleksei Kudrin: Moi Simpatii Na Storone "Edinoi Rossii" ', Rossiiskii Vneshnepoliticheskii Sait. Online. Available at http://www.rvps.ru/r_polit.php?id=952 (accessed 24 March 2011).

Kuzio, T. (2006) 'Civil Society, Youth and Societal Mobilization in Democratic Revolutions', *Communist and Post-Communist Studies*, 39 (3): 365–86.

Kvistad, G. (1999) 'Building Democracy and Changing Institutions: The Professional Civil Service and Political Parties in the Federal Republic of Germany', in J. Brady, B. Crawford and S. Wiliarty (eds), *The Post-War Transformation of Germany: Democracy, Prosperity and Nationhood*, Ann Arbor: University of Michigan Press, pp. 63–93.

Kynev, A. (2007) 'Analiz Partiinykh Spiskov', Carnegie Monitoring. Online. Available at http://monitoring.carnegie.ru/2007/11/2007-duma-elections/analysis-of-kandidate-lists (accessed 11 June 2008).

Laakso, M. and R. Taagepera (1979) 'Effective Number of Parties: A Measure with Application to West Europe', *Comparative Political Studies*, 12 (1): 3–27.

LaPalombara, J. and M. Weiner (1966) 'The Origin and Development of Political Parties', in J. LaPalombara and M. Weiner (eds), *Political Parties and Political Development*, Princeton, NJ: Princeton University Press, pp. 3–42.

Latynina, Yu. (2000) 'Ne v Den'gakh Schast'e', *Sovershenno Sekretno*, 4 (131). Online. Available at http://www.sovsekretno.ru/magazines/article/464 (accessed 24 March 2011).

Lavrenkov, I. and V. Romanov (2007) 'Pis'ma "Edinoi Rossii" Otnesli v Livejournal', *Kommersant*. Online. Available at http://www.kommersant.ru/doc.aspx?Docs ID=828054# (accessed 25 March 2011).

Lawson, F. (2007) 'Intraregime Dynamics, Uncertainty, and the Persistence of Authoritarianism in the Contemporary Arab World', in O. Schlumberger (ed.), *Debating Arab Authoritarianism*, Stanford: Stanford University Press, pp. 109–27.

Ledeneva, A. (2006) *How Russia Really Works*, New York: Cornell University Press.

Leshchenko, N. (2008) 'The National Ideology and the Basis of the Lukashenka Regime in Belarus', *Europe-Asia Studies*, 60 (8): 1419–33.

Levada Center (2007) 'Voting Behaviour – 2007 State Duma Campaign'. Online. Available at http://www.russiavotes.org/duma/duma_vote_2003_listparty.php (accessed 23 March 2011).

Levada Center (2008) 'Uroven' Odobreniya Deyatel'nosti V. Putina Ha Postu Prezidenta Rossii'. Online. Available at http://www.levada.ru/prez07.html (accessed 25 March 2011).

Levchenko, A. (2007) 'Surkov Meets with United Russia Leaders on Rallies Law, Duma Elections', Johnson's Russia List 2007-#17. Online. Available at http://www.cdi.org/russia/johnson (accessed 25 January 2007).

Levitsky, S. (2003, April) 'Autocracy by Democratic Rules: The Dynamics of Competitive Authoritarianism in the Post-Cold War Era', paper presented at the Columbia University Conference on Mapping the Great Zone: Clientelism and the Boundary between Democratic and Democratizing, New York.

Levitsky, S. (2007) 'From Populism to Cleintelism? The Transformation of Labour-Based Party Linkages in Latin America', in H. Kitschelt and S. Wilkinson (eds), *Patrons, Clients and Policies*, Cambridge: Cambridge University Press, pp. 206–26.

Levitsky, S. and L. Way (2010) *Competitive Authoritarianism: Hybrid Regimes after the Cold War*, New York: Cambridge University Press.

Lewis, P. (2000) *Political Parties in Post-Communist Eastern Europe*, London: Routledge.

Lewis, P. (2001) 'The "Third Wave" of Democracy in Eastern Europe: Comparative Perspective on Party Roles and Political Development', *Party Politics*, 7 (5): 543–65.

Lijphart, A. (1971) 'Comparative Politics and the Comparative Method', *American Political Science Review*, 65 (3): 682–93.

Likhtenshtein, A. (2002, March) 'Zakon o Politicheskikh Partiyakh. Strategii Partiinogo Stroitel'stva Rossiiskikh Elit: "Partii Vlasti" ', paper presented at the St Petersburg Institute for the Development of Electoral Systems Conference on Elections in the Russian Federation.

Likhtenshtein, A. (2003) 'Institutsional'nye Usloviya Vozniknoveniya i Funkstionirovaniya "Partii Vlasti" v Rossii i Ukraine: Sravnitel'nye Analiz', unpublished thesis, Moscow State Institute of International Relations.

Lilleker, D. (2003) 'Interviewing the Political Elite: Navigating a Potential Minefield', *Politics*, 23 (3): 207–14.

Lindberg, S. (2006) 'Tragic Protest: Why Do Opposition Parties Boycott Elections?', in A. Schedler (ed.), *Electoral Authoritarianism*, Boulder, CO: Lynn Rienner, pp. 149–63.

Linz, J. (1970) 'An Authoritarian Regime: Spain', in E. Allardt and S. Rokkan (eds), *Mass Politics*, London: Collier-Macmillan, pp. 251–83.

Linz, J. (2000) *Totalitarian and Authoritarian Regimes*, Boulder, CO: Lynn Rienner.

Lipset, S. and S. Rokkan (1967) *Party Systems and Voter Alignments: Cross National Perspectives*, London: Collier-Macmillan.

Little, D. (1991) *Varieties of Social Explanation: An Introduction to the Philosophy of Social Science*, Boulder, CO: Westview Press.

Loginov, A. (2007) Representative for the Government in the State Duma, interview with author, 10 June, Moscow.

Lust-Okar, E. (2007) 'The Management of Opposition: Formal Structures of Contestation and Informal Political Manipulation in Egypt, Jordan and Morocco', in O. Schlumberger (ed.), *Debating Arab Authoritarianism*, Stanford: Stanford University Press, pp. 39–58.

Luzhkov, Y. (2005) *Razvitiye Kapitalizma v Rossii 100 Let Spustya*, Moscow: Moskovkie Uchebniki i Kartolotografiya.

L'vov, S. (2007) Head of Political Research, VTsIOM, interview with author, 25 October, Moscow.

Lynch, A. (2005) *How Russia Is Not Ruled – Reflections on Russian Political Development*, Cambridge: Cambridge University Press.

Lysenko, V. (2007) President of the Institute of Contemporary Politics, interview with author, 4 April, Moscow.

McCoy, J. and D. Myers (eds) (2006) *The Unravelling of Representative Democracy in Venezuela*, Baltimore: Johns Hopkins University Press.

McFaul, M. (1998) 'Russia's Choice: The Perils of Revolutionary Democracy', in T. Colton and J. Hough (eds), *Growing Pains: Russian Democracy and the Election of 1993*, Washington, D.C: Brookings Institution Press, pp. 115–39.

McFaul, M. (1999) 'The Perils of a Protracted Transition', *Journal of Democracy*, 10 (2): 4–8.

McFaul, M. (2001) 'Explaining Party Formation and Non-Formation in Russia – Actors, Institutions and Chance', *Comparative Political Studies*, 34 (10): 1159–87.

McFaul, M. (2002) 'The Fourth Wave of Democracy and Dictatorship: Noncooperative Transitions in the Postcommunist World', *World Politics*, 54 (2): 212–44.

McFaul, M. and S. Markov (1993) *The Troubled Birth of Russian Democracy: Parties, Personalities and Programs*, Stanford: Hoover Institution Press.

Magaloni, B. (2006) *Voting for Autocracy: Hegemonic Party Survival and Its Demise in Mexico.* New York: Cambridge University Press.

Mainwaring, S. (1999) *Rethinking Party Systems in the Third Wave of Democratization: The Case of Brazil*, Stanford: Stanford University Press.

Mainwaring, S. and M. Torcal (2005) 'Party System Institutionalization and Party System Theory after the Third Wave of Democratization', Helen Kellogg Institute for International Studies. Online. Available at http://www.ciaonet.org/wps/klg001/index.html (accessed 24 March 2011).

Mair, P. (1979) 'The Autonomy of the Political: The Development of the Irish Party System', *Comparative Politics*, 11 (4): 445–65.

Makarenko, B. (2000) 'Blok 'Otechestvo – Vsya Rossiya', in M. McFaul, N. Petrov and A. Ryabov (eds), *Rossiya v Izbiratel'nom Tsikle 1999–2000 godov*, Moscow: Moscow Carnegie Centre, pp. 120–39.

Makarkin, A. (2000) 'Partii Vlasti', in M. McFaul, N. Petrov and A. Ryabov (eds), *Rossiya v Izbiratel'nom Tsikle 1999–2000 godov*, Moscow: Moscow Carnegie Centre, pp. 144–54.

Maramatsu, M. and E. Krauss (1990) 'The Dominant Party and Social Coalitions in Japan', in T. Pempel (ed.), *Uncommon Democracies: The One-Party Dominant Regimes*, Ithaca, NY: Cornell University Press, pp. 282–305.

March, L. (2002) *The Communist Party in Post-Soviet Russia*, Manchester: Manchester University Press.

March, L. (2009) 'Managing Opposition in a Hybrid Regime: Just Russia and Parastatal Opposition', *Slavic Review*, 68 (3): 504–27.

Markov, S. (2007) Director of the Institute for Political Studies, interview with author, 16 March, Moscow.

Matveeva, A. (2007) Journalist, *RBK Daily*, interview with author, 2 March, Moscow.

Maximov, V. (2007) Journalist, *Gazeta*, interview with author, 18 March, Moscow.

May, T. (2001) *Social Research: Issues, Methods and Process*, Buckingham: Open University Press.

Medialogia (2007) 'Martovskaya Ottepel' v Efire'. Online. Available at http://www.ng.ru/politics/2007–04–03/4_march.html (accessed 23 March 2011).

Medinskii, V. (2007) United Russia deputy, State Duma, interview with author, 13 April, Moscow.

Medinskii, V. (2011) 'Pust' Leniny Zemlya Budet Pukhom', *Komsomol'skaya Prava*. Online. Available at http://www.kp.ru/daily/25624.4/791039 (accessed 24 March 2011).

Medvedev, D. (2006) 'Dlya Protsvetaniya Vsekh Nado Uchityvat' interesy kazhdogo', *Ekspert*. Online. Available at http://expert.ru/expert/2006/28/interview_medvedev/ (accessed 28 August 2011).

Medvedev, D. (2008a) 'Vystuplenie na IX S"Ezde Partii "Edinoi Rossii" ', Prezident Rossii. Online. Available at http://www.rost.ru/medvedev/report-15–04.html (accessed 25 March 2011).

Medvedev, D. (2008b) 'Poslanie Federal'nomu Sobraniyu Rossiiskoi Federatsii', Prezident Rossii. Online. Available at http://kremlin.ru/appears/2008/11/05/1349_type 63372type63374type63381type82634_208749.shtml (accessed 23 March 2011).

Medvedev, D. (2009) 'Rossiya, vpered! Stat'ya Dmitriya Medvedva'. Online. Available at http://www.kremlin.ru/news/5413 (accessed 24 March 2011).

Mekhanik, A. (2007) Representative of the Club 4 November, interview with author, 8 June, Moscow.

Mereu, F. (2003) 'Kremlin is Betting on United Russia', *The Moscow Times*, 23 September, p. 4.

Merkl, P. (1999) 'The German Response to the Challenge of Extremist Parties, 1949–1994', in J. Brady, B. Crawford and S. Wiliarty (eds), *The Post-War Transformation of Germany: Democracy, Prosperity and Nationhood*, Ann Arbor: University of Michigan Press, pp. 35–62.

Metel'skii, A. (2007) United Russia Deputy, Moscow Legislative Assembly, interview with author, 26 July, Moscow.

Ministerstvo Ustitsii (2011a) 'Perechen' Regional'nykh Otdelenii'. Online. Available at http://www.minjust.ru/ru/activity/nko/partii/ER/ (accessed 24 March 2011).

Ministerstvo Ustitsii (2011b) 'Federal'nyi Spisok Ekstremistskikh Materialov'. Online. Available at http://www.minjust.ru/ru/activity/nko/fedspisok (accessed 24 March 2011).

Ministerstvo Ustitsii (2011c) 'Perechen'. Online. Available at http://www.minjust.ru/ru/activity/nko/perechen (accessed 24 March 2011).

Mitrokhin, S. (2007) Yabloko deputy, Moscow Legislative Assembly, interview with author, 17 October, Moscow.

Molodaya Gvardiya (2009a) 'Ustav Organizatsii'. Online. Available at http://www.molgvardia.ru/statutes (accessed 5 June 2009).

Molodaya Gvardiya (2009b) 'Programma Molodezhnoi Praimariz v Voprosakh i Otvetakh'. Online. Available at http://www.er34.ru/mg/4.htm (accessed 10 June 2009).

Morozov, O. (2006) 'Tsentrizm – Eto Antiradikalizm', Edinaya Rossiya. Online. Available at http://www.edinros.ru/news.html?id=112414 (accessed 23 April 2007.

Moser, D. (1998) 'Sverdlovsk: Mixed Results in a Hotbed of Regional Autonomy', in T. Colton and J. Hough (eds), *Growing Pains: Russian Democracy and the Election of 1993*, Washington, D.C.: Brookings Institution Press, pp. 141–76.

Moser, R. (1999) 'Independents and Party Formation: Elite Partisanship as an Intervening Variable', *Comparative Politics*, 31 (2): 147–65.

Moshkin, M. and I. Romanov (2007) 'General'naya Imitatsiya Vybora', *Nezavisimaya gazeta*. Online. Available at http://www.ng.ru/politics/2007–03–12/1_veshnyakov.html (accessed 25 March 2011).

Moskvin-Tarkhanov, M. (2007) United Russia deputy, Moscow Legislative Assembly, interview with author, 7 June, Moscow.

Mukhin, A. (2006) *Federal'naya i Regional'naya Elita Rossii 2005–2006*, Moscow: Tsentr Politicheskoi Informatsii.

Muller, W. (1997) 'Inside the Black Box: A Confrontation of Party Executive Behaviour and Theories of Party Organisational Change', *Party Politics*, 3 (3): 293–313.

Munck, G. (2001) 'Review Article: The Regime Question – Theory Building in Democracy Studies', *World Politics*, 54: 119–44.

Munck, G. (2006) 'Drawing Boundaries: How to Craft Intermediate Regime Categories', in A. Schedler (ed.), *Electoral Authoritarianism*, London: Lynne Rienner, pp. 27–40.

Nagornykh, I. (2007) 'Edinaya Rossiya Otkazalas' ot Debatov', *Kommersant*. Online. Available at http://www.kommersant.ru/doc.aspx?DocsID=820209 (accessed 23 March 2011).

Nagornykh, I. (2009a) 'Edinaya Rossiya Osvaivaet Nulevoe Chtenie', *Kommersant*. Online. Available at http://www.kommersant.ru/doc.aspx?fromsearch=9e014f68–5849–4f08–89ab-e918e1208a83&docsid=1159532 (accessed 25 March 2011).

Nagornykh, I. (2009b) 'Prezident Vstretitsya s Budushchimi Vydvizhenstami', *Kommersant*. Online. Available at http://www.kommersant.ru/doc.aspx? DocsID=1129193 (accessed 25 March 2011).

Needleman, C. and M. Needleman (1969) 'Who Rules Mexico: A Critique of Some Current Views on the Mexican Political Process', *Journal of Politics*, 31: 1011–34.

Nemtsov, B. (2007) 'URF member', interview with author, 26 June, Moscow.

NIC (2008) 'Global Trends 2025: A Transformed World', National Intelligence Council. Online. Available at http://www.dni.gov/nic/NIC_2025_project.html (accessed 25 March 2011).

Nikiforova, M. (2002) 'Kuda Partiya Poshlet', *Vremya*. Online. Available at http://www.vremya.ru/2002/102/4/23952.html (accessed 25 March 2011).

Nikonov, V. (1993) 'Tri Konstitutsionnye Bomby, Kotorye Mogut Vzorvat'sya', in V. Nikonov (ed.), *Epokha Peremen: Rossiya 90-Kh Glazami Konservatora*, Moscow: Yazyki Russkoii Kul'tury, pp. 141–4.

Nikonov, V. (2007) President of the Polity Foundation, interview with author, 28 June, Moscow.

Nikonov, V. and S. Shakhrai (1994) 'Konservativnyi Manifest', in V. Nikonov (ed.), *Epokha Peremen: Rossiya 90-kh Glazami Konservatora*, Moscow: Yazyki Russkoi Kul'tury, pp. 11–28.

North, D. (1990) *Institutions, Institutional Change and Economic Performance*, Cambridge: Cambridge University Press.

Novye Region (2009) 'Skandal: Zhitel' Sverdlovskoi Oblasti isklyuchen iz Kadrovogo Reserva Medvedeva'. Online. Available at http://nr2.ru/ekb/221469.html (accessed 25 March 2011).

Oates, S. (2000, April) 'The Advent of the Broadcast Party: Parties, Voters and Television in Russia 1993–1999', paper presented at the Political Studies Association–UK 50th Annual Conference, London.

Oates, S. (2006) *Television, Democracy and Elections in Russia*, Abingdon: Routledge.

Obshchestvennaya Palata (2010) 'Chleny Obshchestvennoi Palaty'. Online. Available at http://www.oprf.ru/ru/chambermembers/members2008/ (accessed 25 March 2011).

ODIHR (2001) 'Republic of Belarus Presidential Election 9 September 2001', OSCE/ODIHR Election Observation Mission Report. Online. Available at http://www.osce.org/odihr/elections/belarus/14459 (accessed 25 March 2011).

ODIHR (2003) 'Parliamentary Elections in Russian Federation – Preliminary Media Monitoring'. Online. Available at http://www.memo98.sk/en/index.php?base=data/foreign/rus/russia_parl_2003.txt (accessed 24 March 2011).

ODIHR (2004) 'Russian Federation Elections to the State Duma 7 December 2003', OSCE/ODIHR Election Observation Mission Report. Online. Available at http://www.osce.org/documents/odihr/2004/01/1947_en.pdf (accessed 3 February 2006).

ODIHR (2004) 'Republic of Belarus Parliamentary Elections 17 October 2004', OSCE/ODIHR Election Observation Mission Report. Online. Available at http://www.osce.org/odihr/elections/belarus/38658 (accessed 25 March 2011).

ODIHR (2006) 'Republic of Belarus Presidential Election 19 March 2006', OSCE/ODIHR Election Observation Mission Report. Online. Available at http://www.osce.org/odihr/elections/belarus/19395 (accessed 25 March 2011).

O'Donnell, G. (1973) *Modernization and Bureaucratic Authoritarianism*, Berkley: University of California Press.

O'Donnell, G. (1994) 'Delegative Democracy', *Journal of Democracy*, 5 (1): 55–69.

Olenich, N. (2006) Zakony Privlekatel'nosti, *Gazeta.* Online. Available at http://www.gazeta.ru/comments/2006/03/07_a_558019.shtm (accessed 15 March 2006).

Orlov, D. (2011) '100 Vedushchikh Politikov v Rossii v 2010 godu', *Nezavisimaya gazeta.* Online. Available at http://www.ng.ru/ideas/2011–01–12/9_top100.html (accessed 24 March 2011).

Orlova, N. (2007) Chair of the Federal Political Council, Molodaya Gvardiya, interview with author, 18 June, Moscow.

Osborn, A. (2004) 'CIA Angers Russia by Predicting Break-up of State within 10 Years', *The Independent.* Online. Available at http://www.independent.co.uk/news/world/europe/cia-angers-russia-by-predicting-breakup-of-state-within-10-years-561727.html (accessed 24 March 2011).

Osipova, G. and V. Lokosova (2007) *Rossiya Predposylki Preodoleniya Sistemnogo Krizisa*, Moscow: Institut Sotsialno-Politicheskikh Issledovanii RAN.

Ostrovsky, A. and N. Buckley (2005) 'Russian PM Snubbed on Spending', *Financial Times.* Online. Available at http://www.ft.com/cms/s/0/f91882f2–556a-11da-8a74–00000e25118c.html?nclick_check=1 (accessed 25 November 2008).

Otechestvo Vsya-Rossiya (1999) 'Predvybornaya Progamma'. Online. Available at http://www.panorama.ru/works/vybory/party/p-ovr.html (accessed 28 August 2011).

Ottaway, M. (2003) *Democracy Challenged: The Rise of Semi-Authoritarianism.* Washington, D.C.: Carnegie Endowment for International Peace.

Ovchinnikov, B. (2001) 'Parlamentskie Vybory 1999: Statisticheskie Anomalii', in N. Petrov (ed.), *Regiony Rossii v 1999: Ezhegodnoe Prilozhenie k Politicheskomu Al'manakhu Rossii*, Moscow: Moscow Carnegie Centre, pp. 225–37.

Overchenko, M. (2009) 'Khodorkovskii Vinoven v Finansirovanii KPRF', *Vedomosti*. Online. Available at http://www.vedomosti.ru/newsline/news/2009/07/21/805978 (accessed 25 March 2011).

Oversloot, H. and R. Verheul (2000) 'The Party of Power in Russian Politics', *Acta Politica*, 35: 123–45.

Oversloot, H. and R. Verheul (2006) 'Managing Democracy: Political Parties and the State in Russia', *Journal of Communist Studies and Transition Politics*, 22 (3): 383–405.

Panebianco, A. (1988) *Political Parties: Organisation and Power*, Cambridge: Cambridge University Press.

Panfilov, O. (2007) Head of the Centre of Extreme Journalism, interview with author, 13 April, Moscow.

Panorama (1999) 'Kandidaty po Obshchefederal'nomu Okrugy ot NDR', Informatsionno-Issledovatel'skii Tsentr 'Panorama'. Online. Available at http://www.panorama.ru/works/vybory/party/ndr2.html (accessed 23 March 2011).

Panorama (2007) 'Est' Takie Partii', Informatsionno-Issledovatel'skii Tsentr 'Panorama'. Online. Available at http://www.scilla.ru/works/partii07/index.html (accessed 23 March 2011).

Pashentsev, E. (2007) Professor of History, Moscow State Institute for International Relations (MGIMO), interview with author, 9 April, Moscow.

Pel'ts, A. (1993) 'Demokraticheskie Sili Ob'edinilis' ', *Krasnaya zvezda*. Online. Available at http://dlib.eastview.com/sources/article.jsp?id=3363887 (accessed 7 January 2009).

Pempel, T. (ed.) (1990a) *Uncommon Democracies: The One-Party Dominant Regimes*, Ithaca, NY: Cornell University Press.

Pempel, T. (1990b) 'Introduction. Uncommon Democracies: The One-Party Dominant Regimes', in T. Pempel (ed.), *Uncommon Democracies: The One-Party Dominant Regimes*, Ithaca, NY: Cornell University Press, pp. 1–32.

Perepis' (2002) 'Nasilenie po Natsional'nosti i Vladeniyu Russkim Yazykom po Sub'ektam Rossiiskoi Federatsii', Vcerossiiskaya Perepis' Naseleniya 2002 goda'. Online. Available at http://www.perepis2002.ru/ct/doc/TOM_04_03.xls (accessed 25 March 2011).

Petrov, N. (2003) 'Russia's "Party of Power" Takes Shape', *Russian and Eurasia Review*, 11 (16): 8–10.

Petrova, A. (1999) 'Why is NATO Bombing Yugoslavia?', FOM. Online. Available at http://bd.english.fom.ru/report/map/eof991703 (accessed 14 February 2010).

Petrova, A (2004) 'How Much Did the Media Influence the Vote?', FOM. Online. Available at http://bd.english.fom.ru/report/cat/societas/mass_media/election_feature/eof 041104 (accessed 1 October 2008).

Pierson, P. (1993) 'When Effect Becomes Cause: Policy Feedback and Political Change', *World Politics*, 45: 595–628.

Pierson, P. (2004) *Politics in Time: History, Institutions, and Social Analysis*, Princeton, NJ: Princeton University Press.

Pietilainen, J. (2008) 'Media Use in Putin's Russia', *Journal of Communist Studies and Transition Politics*, 24 (3): 365–85.

Ponomarev, L. (2007) Director of the All-Russian Movement for Human Rights, interview with author, 19 October, Moscow.

Popov, S. (2007) United Russia deputy, State Duma, interview with author, 7 March, Moscow.

Pozdnyakova, E. (2009) 'Edinuyu Rossiyu Lishili Illyuzii', Tsentr Politicheskikh Tekhnologii. Online. Available at http://www.politcom.ru/article.php?id=7971 (accessed 25 March 2011).

Pribylovskii, V. (1992) *Dictionary of Political Parties and Organisations in Russia*, Washington, D.C.: Centre for Strategic and International Studies.

Pribylovskii, V. (2006) 'Pyat Bashen. Politicheskaya Topografiya Kremlya'. Online. Available at http://www.anticompromat.ru/putin/5bashen.html (accessed 3 March 2008).

Pribylovskii, V. (2007) Director of the Panorama think tank, interview with author, 17 February, Moscow.

Purcell, S. (1973) 'Decision Making in an Authoritarian Regime: Theoretical Considerations from a Mexican Case Study', *World Politics*, 26 (1): 28–54.

Putin, V. (2001) 'Poslanie Federal'nomu Sobraniyu Rossiiskoi Federatsii', Prezident Rossii. Online. Available at http://www.kremlin.ru/sdocs/appears.shtml?date_to=2008/05/06&stype=63372 (accessed 3 May 2009).

Putin, V. (2007a) 'Poslanie Federal'nomu Sobraniyu Rossiiskoi Federatsii', Prezident Rossii. Online. Available at http://archive.kremlin.ru/text/appears/2007/04/125339. Shtml (accessed 23 March 2011).

Putin, V. (2007b) 'Putin Schitaet "Edinuyu Rossiyu" Zalogom Deesposobnosti Parlamenta', Gosudarstvennaya Duma. Available at http://gduma.ru/news_25.htm (accessed 25 March 2011).

Radkevich, S. (2007) Head of Analytical Department, 'Niccolo M', interview with author, 20 April, Moscow.

Rae, D. (1971) *The Political Consequences of Electoral Laws*, New Haven, CT: Yale University Press.

Ragin, C. (1987) *The Comparative Method*, Berkley: University of California Press.

Randall, V. and L. Svasand (2002) 'Party Institutionalization in New Democracies', *Party Politics*, 8 (1): 5–29.

Rasmussen, T. (1969) 'Political Competition and One-Party Dominance in Zambia', *Journal of Modern African Studies*, 7 (3): 407–24.

Reddy, T. (2006) 'INC and ANC: A Comparative Analysis', in A. Ostheimer (ed.), *Challenges to Democracy by One-Party Dominance: A Comparative Assessment*, Johannesburg: Konrad-Adenaur-Stiftung, pp. 55–61.

Reding, A. (1991) 'Mexico: The Crumbling of the "Perfect Dictatorship" ', *World Policy Journal*, 8 (2): 255–84.

Reeve, D. (1985) *Golkar of Indonesia: An Alternative to the Party System*, Oxford: Oxford University Press.

Remington, T. (2001) *The Russian Parliament: Institutional Evolution in a Transitional Regime, 1989–1999*, Chelsea: Sheridan Books.

Remington, T. (2003) 'Prospects for a Democratic Left in Post-Communist Russia', *Journal of Policy History*, 15 (1): 130–48.

Remington, T. (2005) 'Putin, the Duma and Political Parties', in D. Herspring (ed.), *Putin's Russia: Past Imperfect, Future Uncertain*, Oxford: Rowman & Littlefield, pp. 31–51.

Remington, T. (2006) 'Presidential Support in the Russian State Duma', *Legislative Studies Quarterly*, 31 (1): 5–32.

Remington, T. and S. Smith, (1995) 'The Development of Parliamentary Parties in Russia', *Legislative Studies Quarterly*, 20: 457–89.

Reuter, O. (2010) 'The Politics of Dominant Party Formation: United Russia and Russia's Governors', *Europe-Asia Studies*, 62 (2): 293–7.

Reuter, O. and T. Remington (2009) 'Dominant-Party Regimes and the Commitment Problem: The Case of United Russia', *Comparative Political Studies*, 42: 501–25.

Richards, D. (1996) 'Elite Interviewing: Approaches and Pitfalls', *Politics*, 16 (3): 199–204.

Riggs, J. and P. Schraeder (2005) 'Russia's Political Party System as a (Continued) Impediment to Democratization: The 2003 Duma and 2004 Presidential Elections in Perspective', *Demokratizatsiya*, 13 (1): 141–51.

Riker, W. (1988) *Liberalism against Populism*, Prospect Heights, Il: Waveland Press.

Roberts, C. and T. Sherlock (1999) 'Bringing the Russian State Back in: Explanations of the Derailed Transition to Market Democracy', *Comparative Politics*, 31 (4): 477–98.

Roberts, K. (2006) 'Do Parties Matter? Lessons from the Fujimori Experience', in J. Carrion (ed.), *The Fujimori Legacy: The Rise of Electoral Authoritarianism in Peru*, University Park: Pennsylvania State University Press, pp. 81–101.

Rodan, G. (2004) *Transparency and Authoritarian Rule in South East Asia: Singapore and Malaysia*, London: RoutledgeCurzon.

Rodin, I. (1997) 'Meditsinskii Impichment', *Nezavisimaya gazeta*. Online. Available at http://dlib.eastview.com/sources/article.jsp?id=297269 (accessed 2 June 2008).

Rodin, I. (2010) 'Edinrossov Ne Ponyali', *Nezavisimaya gazeta*. Online. Available at http://www.ng.ru/politics/2010–12–14/2_edro.html (accessed 25 March 2011).

Rodin, I. and A. Samarina (2009a) 'Deputatskii Demarsh', *Nezavisimaya gazeta*. Online. Available at http://www.ng.ru/politics/2009–10–15/1_demarsh.html (accessed 24 March 2011).

Rodin, I. and A. Samarina (2009b) 'Po Kreslu na Partiyu', *Nezavisimaya gazeta*. Online. Available at http://www.ng.ru/politics/2009–05–28/1_chair.html (accessed March 23 2011).

Rokkan, S. (1975) 'Dimensions of State Formation and Nation Building: A Possible Paradigm for Research on Variations within Europe', in C. Tilly (ed.), *The Formation of National States in Western Europe*, Princeton, NJ: Princeton University Press, pp. 562–600.

Rose, R. (1969) 'The Variability of Party Government: A Theoretical and Empirical Critique', *Political Studies*, 17 (4): 413–45.

Rose, R. (2001) *Russian under Putin, Vol. 350 New Russia Barometer*, Aberdeen: CSPP Publications.

Rose, R. and D. Shin (2001) 'Democratization Backwards: The Problem of Third-Wave Democracies', *British Journal of Political Science*, 31: 331–54.

Rose, R. and N. Munro (2002) *Elections without Order: Russia's Challenge to Vladimir Putin*, Cambridge: Cambridge University Press.

Rose, R., W. Mishler and C. Haerpfer (1998) *Democracy and Its Alternatives: Understanding Post-Communist Societies*, Baltimore: Johns Hopkins University Press.

Rosstat (2009) 'Naseleniya Sub'ektov Rossiiskoi Federatsii', Federal'naya Sluzhba Gosudarstvennoi Statistiki. Online. Available at http://www.gks.ru/bgd/regl/B09_16/IssWWW.exe/Stg/01–07.htm (accessed 23 March 2011).

Rost (2010) 'Sostav Presiduma Soveta', Prioritetnye Natsional'nye Proekty. Online. Available at http://www.rost.ru/main/sovet/sovet_3/asovet_3.shtml (accessed 25 March 2011).

Round Table (2007) 'Plany Partii – Plany Naroda?', Politklub 'Izvestii', 29 March, Moscow.

Russia Votes (n.d.) 'Results of Previous Elections to the Russian State Duma'. Online. Available at http://www.russiavotes.org (accessed 10 January 2011).

'Transitions to Democracy: Toward a Dynamic Model', *Comparative Politics*, 20(3), 337–63.

Ryabov, A. (1998) 'Partiya Vlasti v Politicheskoi Sisteme Sovremennoi Rossii', in M. McFaul, S. Markov and A. Ryabov (eds), *Formirovanie Partiino-Politicheskoi Sistemy v Rossii*, Moscow: Moscow Carnegie Centre, pp. 80–96.

Ryzhkov, V. (2007) Independent deputy, State Duma, interview with author, 12 April, Moscow.

Ryzhkov, V. (2011) '21 Vek Mozhet Udvoit' Kolichestvo Gosudarstv v Mire', interview with *Rossiiskaya gazeta*, Online. Available at http://www.ryzkov.ru/pg.php?id=8635 (accessed 25 March 2011).

Sadchikov, A. (2001) 'Nedovotum: Doveriya Pravitel'stvu Ne Dobavilos', *Izvestiya*. Online. Available at http://dlib.eastview.com/sources/article.jsp?id=3056405 (accessed 3 January 2009).

Sadchikov, A. (2003) 'Partiinyi Destabilizator', *Izvestiya*. Online. Available at http://www. izvestia.ru/russia/article30469/ (accessed 25 March 2011).

Sadkovskaya, T. (1998) '7 Dnei: iz "Belogo Doma" – v "Nash Dom" ', *Rossiiskie Vesti*. Online. Available at http://dlib.eastview.com/sources/article.jsp?id=1920123 (accessed 7 January 2009).

Sakwa, R. (1997) 'The Regime System in Russia', *Contemporary Politics*, 3: 7–25.

Sakwa, R. (2005) 'The 2003–2004 Russian Elections and Prospects for Democracy', *Europe-Asia Studies*, 57 (3): 369–98.

Sakwa, R. (2010) *The Crisis or Russian Democracy: The Dual State, Factionalism and the Medvedev Succession*, Cambridge: Cambridge University Press.

Samarina, A. (2009) 'Edinaya Rossiya Radvoditsya c "Delovoi . . ." ', *Nezavisimaya gazeta*. Online. Available at http://www.ng.ru/politics/2009–06–29/3_edro.html (accessed 25 March 2011).

Samarina, A. and R. Rodin (2010) 'Partiino-Politicheskii Modern', *Nezavisimaya gazeta*. Online. Available at http://www.ng.ru/politics/2010-04-07/1_modern.html (accessed 23 March 2011).

Samoilova, S. (2009) 'Zyuganov Na Svyazi', Tsentr Politicheskikh Tekhnologii. Online. Available at http://www.politcom.ru/7809.html (accessed 21 March 2011).

Samuels, D. (2002) 'Presidentialized Parties: The Separation of Powers and Party Organisation and Behaviour', *Comparative Political Studies*, 35 (4): 461–83.

Sartori, G. (1976) *Parties and Party Systems: A Framework for Analysis*, Colchester: ECPR Press. Reprinted 2005.

Satarov, G. (2004) 'The New Duma 2004. Consequences for Business?', Round Table, European Business Club. Online. Available at http://www.aebrus.ru/files/File/EventFiles/Other_Events/20040119/19_01_2004_Summary_eng.doc (accessed 23 March 2011).

Satarov, G. (2006) *Biznes i Korruptsiya: Problemy Protivodeistviya*, Moscow: INDEM.

Satarov, G. (2007) Head of the INDEM Foundation, interview with author, 16 July, Moscow.

Satter, D. (2003) *Darkness at Dawn: The Rise of the Russian Criminal State*, New Haven, CT: Yale University Press.

Savin, V. (2007) Head of Workers and Trade Unions, CPRF, interview with author, 23 November, Moscow.

Savvateeva, I. (1995) 'Nash Dom Vzglyad iznutri', *Izvestiya*. Online. Available at http://dlib. eastview.com/sources/article.jsp?id=3185136 (accessed 5 December 2008).

Schattschneider, E. (1942) *Party Government*, New York: Farrar and Rinehart.

Schedler, A. (2002) 'The Menu of Manipulation', *Journal of Democracy*, 13 (2): 36–50.

Schedler, A. (eds) (2006a) *Electoral Authoritarianism*, London: Lynne Rienner Publishers.

Schedler, A. (2006b) 'The Logic of Electoral Authoritarianism', in A. Schedler (ed.), *Electoral Authoritarianism*, London: Lynne Rienner, pp. 1–23.

Scott, R. (1959) *Mexican Government in Transition*, Urbana: University of Illinois Press.

Segedinenko, N. (2004) *Chetvertaya Gosudarstvennaya Duma: Anatomicheskii Atlas*, Moscow: Tsentr Politicheskoi Informatsii.

Seshia, S. (1998) 'Divide and Rule in Indian Party Politics: The Rise of the Bharatiya Janata Party', *Asian Survey*, 38 (11): 1036–50.

Shalev, M. (1990) 'The Political Economy of Labour Party Dominance and Decline in Israel', in T. Pempel (ed.), *Uncommon Democracies*, Ithaca, NY: Cornell University Press, pp. 83–127.

Shefter, M. (1994) *Political Parties and the State – the American Historical Experience*, Princeton, NJ: Princeton University Press.

Shelishch, P. (2007) State Duma deputy, United Russia faction, interview with author, 21 March, Moscow.

Shkel', T. (2008) 'Dialog po-Profsoyuznomu', *Rossisskaya gazeta*. Online. Available at http://m.rg.ru/2008/04/11/ed-rossiya.html (accessed 25 March 2011).

Shevtsova, L. (2001) 'From Yeltsin to Putin: The Evolution of Presidential Power', in A. Brown and L. Shevtsova (eds), *Gorbachev, Yeltsin, Putin: Political Leadership in Russia's Transition*, Washington, D.C: Carnegie Endowment for International Peace, pp. 67–112.

Shevtsova, L. (2005) 'Vladimir Putin's Political Choice: Towards Bureaucratic Authoritarianism', in A. Pravda (ed.), *Leading Russia: Putin in Perspective*, Oxford: Oxford University Press, pp. 229–53.

Shevtsova, L. (2007) *Russia Lost in Transition*, Washington, D.C: Carnegie Endowment for International Peace.

Shleifer, A. and D. Treisman (2004) 'A Normal Country', *Foreign Affairs*, 83 (2): 20–38.

Shoigu, S. (2006) 'V Pravitel'stve Dolzhny Rabotat' Professionaly', Edinaya Rossiya. Online. Available at http://edinros.er.ru/er/text.shtml?17035/101750 (accessed 3 April 2009).

Shtykina, A., and A. Barakhova (2010) 'Kto v Sovete Federatsii Samii Predstavitel'nye', *Kommersant*. Online. Available at http://www.kommersant.ru/doc.aspx?Docs ID=1368448 (accessed 5 February 2011).

Shvedov, A. (2004) 'Gryzlov Obeshchal Izmenit' Zakon o Mitingakh', *Izvestiya*. Online. Available at http://dlib.eastview.com/sources/article.jsp?id=6155367 (accessed 10 May 2009).

Sidorov, D. (2007) Secretary of the Independent Trade Union, interview with author, 22 November, Moscow.

Sigudkin, A. (2007) United Russia deputy, State Duma, interview with the author, 21 May, Moscow.

Silantiev, A. (2007) United Russia, International Relations Department, interview with author, 19 February, Moscow.

Singh, H. (1998) 'Tradition, UMNO and Political Succession in Malaysia', *Third World Quarterly*, 19 (2): 241–54.

Skocpol, T. (1979) *States and Social Revolutions*, Cambridge: Cambridge University Press.

Skocpol, T. (1985) 'Bringing the State Back In: Strategies of Analysis in Current Research', in P. Evans, D. Rueschemeyer and T. Skocpol (eds), *Bringing the State Back In*, London: Cambridge University Press, pp. 3–37.

Skoryi, R. (2007) Director of International Affairs, Centre for Social Conservative Policy, interview with author, 9 July, Moscow.

Slider, D. (2010) 'How United is United Russia? Regional Sources of Intra-Party Conflict', *Journal of Communist Studies and Transition Politics*, 26 (2): 257–75.

Slomatin, E. (2007) LDPR deputy, State Duma, interview with author, 12 November, Moscow.

Slyusarenko, S. (1999) 'Primakov Obidel Gubernatora', *Kommersant*". Online. Available at http://www.kommersant.ru/doc.aspx?fromsearch=f9027ff6–0acf-4e45–827c-3693cd1f3c58&docsid=213561 (accessed 25 March 2011).

Smirnov, O. (2007) 'Edinaya Rossiya' Partiya Milliarderov', *Sotsial'nyi vopros*, June, p. 4.

Smith, B. (2005) 'Life of the Party: The Origins of Regime Breakdown and Persistence under Single-Party Rule', *World Politics*, 57: 421–51.

Smith, M. (2006) 'Sovereign Democracy: The Ideology of Yedinaya Rossiya', Defence Academy of the United Kingdom Research Centre, Russian Series 06/37. Online. Available at http://www.da.mod.uk/CSRC/documents/Russian/06%2837%29 MAS.pdf (accessed 23 March 2011).

Smyth, R. (2002) 'Building State Capacity from the Inside Out: Parties of Power and the Success of the President's Reform Agenda in Russia', *Politics & Society*, 30 (4): 555–78.

Smyth, R. (2006) *Candidate Strategies and Electoral Competition in the Russian Federation: Democracy without Foundation*, New York: Cambridge University Press.

Smyth, R., A. Lowry and B. Wilkening (2007) 'Engineering Victory: Institutional Reform, Informal Institutions and the Formation of a Hegemonic Party Regime in the Russian Federation', *Post-Soviet Affairs*, 23 (2): 118–37.

Snieder, M. (2007) Union of Right Forces member, interview with author, 16 March, Moscow.

Sova (2009a) 'Pervoaprel'skoe Obnovlenie Spiska Ekstremistskikh Materialov', Sova Informatsionno-Analiticheskii Tsentr. Online. Available at http://xeno.sova-center.ru/89CCE27/89CD1C9/CBC8AA7?view=all (accessed 23 March 2011).

Sova (2009b) 'Kommentarii po Povodu Opredeleniya Verkhovnogo Suda RF po Zhalobe SPS', Sova Informatsionno-Analiticheskii Tsentr. Online. Available at http://xeno.sova-center.ru/89CCE27/89CD253/C61841D (accessed 26 March 2011).

Sovet Federatsii (2011) 'Komitety i Komissii'. Online. Available at http://www.council.gov.ru/committee/index.html (accessed 23 March 2011).

Stanley, A. (1995) 'On TV, Russian Party Is More Equal Than Others', *The New York Times*. Online. Available at http://www.nytimes.com/1995/12/06/world/on-tv-russian-party-is-more-equal-than-others.html?n=Top%2fReference%2fTimes%20Topics%2f Subjects%2fT%2fTelevision (accessed 24 March 2011).

Stepan, A. and C. Skach (1994) 'Presidentialism and Parliamentarianism in Comparative Perspective', in J. Linz and A. Valenzuela (eds), *The Failure of Presidential Democracy: Comparative Perspectives*, vol. 1, Baltimore: Johns Hopkins University Press, pp. 119–36.

Stoev, S. (2010) 'Itogi 2009. s 'Medvezh'im' Rylom Ne Puskayut v Evropeiskii Klub. Edinaya Rossiya Sed'moi god kak Bezuspeshno Vstupaet v Evropeiskuyu Narodnuyu Partiyu, Kommunisticheskaya Partiya Rossiiskoi Federatsii'. Online. Available at http://kprf.ru/opponents/74904.html? (accessed 25 March 2011).

Stokes, D. (1992) 'Valence Politics', in D. Kavanagh (ed.), *Electoral Politics*, Oxford: Clarendon Press, pp. 147–64.

Stoner-Weiss, K. (2006) *Resisting the State: Reform and Retrenchment in Post-Soviet Russia*, Cambridge: Cambridge University Press.

Suleiman, E. (1994) 'Presidentialism and Political Stability in France', in J. Linz and A. Valenzuela (eds), *The Failure of Presidential Democracy*, Baltimore: Johns Hopkins University Press, pp. 137–62.

Surkov, V. (2006a) 'Surkov Obeshchaet Vvesti v Rossii Dvykhpartiinuyu Sistemu', *Nezavisimaya gazeta*. Online. Available at http://news.ng.ru/2006/08/16/1155721835.html (accessed 20 March 2011).

Surkov, V. (2006b) 'Chto Skazal Vladislav Surkov', *Kommersant"*. Online. Available at http://www.kommersant.ru/doc.html?DocID=698018&IssueId=30174 (accessed 25 March 2011).

Svyatenko, N. (2007) United Russia deputy, Moscow Legislative Assembly, interview with author, 4 June, Moscow.

Szczerbiak, A. (1999) 'Testing Party Models in East-Central Europe: Local Party Organization in Post Communist Poland', *Party Politics*, 5 (4): 525–37.

Terent'ev, I. (2010) 'Vlast' I Den'gi – 2010. Reiting Forbes', *Forbes*. Online. Available at http://www.forbes.ru/ekonomika-package/vlast/54942-vlast-i-dengi-2010-reiting-forbes (accessed 27 March 2011).

Thackrah, S. (2000, October) 'Unpacking One Party Dominance', paper presented at the Annual Conference of the Australian Political Studies Association, Canberra.

The Economist (2010) 'Democracy Index 2010: Democracy in Retreat'. Online. Available at http://graphics.eiu.com/PDF/Democracy_Index_2010_web.pdf (accessed 24 March 2011).

Thelen, K. (1999) 'Historical Institutionalism in Comparative Politics', *Annual Review of Political Science*, 2: 369–404.

Tilly, C. (1975) 'Western State-Making and Theories of Political Transformation', in C. Tilly (eds), *The Formation of National States in Western Europe*, Princeton, NJ: Princeton University Press, pp. 601–38.

Tkacheva, A. (2008, May) 'Governors as Poster-Candidates in Russia's Legislative Elections, 2003–2008', paper presented at the Temple University Conference on State Politics and Policy, Philadelphia.

Toole, J. (2003) 'Straddling the East–West Divide: Party Organisation and Communist Legacies in East Central Europe', *Europe-Asia Studies*, 55 (1): 101–18.

Toropov, D. (2005) 'Molodaya Gvardiya Dlya Okkupantov?, *Zavtra*. Online. Available at http://www.zavtra.ru/cgi/veil/data/zavtra/05/628/23.html (accessed 25 March 2011).

Tregubova, E., and O. Petrova (2002) 'Novoe Litso Partii Vlasti', *Kommersant Vlast*. Online. Available at http://www.kommersant.ru/doc.aspx?fromsearch=4f938875–0407–444a-b3c2–6bb5de1b1f2f&docsid=352657 (accessed 25 March 2011).

Treisman, D. (1998a) 'Between the Extremes: Moderate Reform and Centrist Blocs in the 1993 Election', in T. Colton and J. Hough (eds), *Growing Pains: Russian Democracy and the Election of 1993*, Washington, D.C.: Brookings Institution Press, pp. 141–76.

Treisman, D. (1998b) 'Dollars and Democratisation: The Role of Power and Money in Russia's Transitional Elections', *Comparative Politics*, 31 (1): 1–21.

Trekhov, A. (2007) 'Kremlevskii Patrul'. 'Mestnye' i 'Nashi' Pomogut Pravookhranitelyam', *Moi Raion*, 37 (235), p. 2.

Tret'yakov, V. (1999) 'Proshchai, "Yabloko" ', *Nezavisimaya gazeta*. Online. Available at http://dlib.eastview.com/sources/article.jsp?id=323830 (accessed 5 June 2008).

Tret'yakov, V. (2003) 'Tendenstii. Svobodny Li SMI Rossii?', *Rossiskaya gazeta*, 11 September, p. 7.

Tret'yakov, V. (2007) 'Suverennaya Demokratiya', in G. Pavlovskii (ed.), *Pro Suverennuyu Demokratiyu*, Moscow: Evropa, p. 627.

Tsygankov, V. (2003) 'Medvedi' – Samye Bogatye i Ekonomnye', *Nezavisimaya gazeta*. Online. Available at http://www.ng.ru/politics/2003–08–05/2_bears.html (accessed 10 June 2008).

Turovskii, R. (2006) 'Regional'nye Vybory v Rossii: Sluchai Atipichnoi Demokratii', in B. Makarenko and I. Bunin (eds), *Tekhnologii Politiki*, Moscow: Tsentr Politicheskikh Tekhnologii, pp. 143–90.

Tyagunov, A. (2007) United Russia deputy, State Duma, interview with author, 24 October, Moscow.

Umland, A. (2006) 'Zhirinovsky in the First Russian Republic: A Chronology of Events 1991–1993', *Journal of Slavic Military Studies*, 19: 193–241.

Urban, M. (1992) 'Boris El'tsin, Democratic Russia and the Campaign for the Russian Presidency', *Soviet Studies*, 44 (2): 187–207.

Urban, M. and V. Gel'man (1997) 'The Development of Political Parties in Russia', in K. Dawisha and B. Parrott (eds), *Democratic Changes and Authoritarian Reactions in Russia, Ukraine, Belarus, and Moldova*, Cambridge: Cambridge University Press, pp. 175–219.

van Biezen, I. (2005) 'On the Theory and Practice of Party Formation and Adaptation in New Democracies', *European Journal of Political Research*, 44: 147–74.

Vasil'tsov, S (2007) ' "Edinaya Rossiya": Popytki Prevratit'sya Partii dlya Vlasti' v Partiyu Vlasti' ', Tsentr Issledovanii Politicheskoi Kul'turii Rossii. Online. Available at http://cipkr.ru/research/ind/_08042007vlast.html (accessed 25 March 2011).

Vedomosti (2010) 'Ne Nado Putina'. Online. Available at http://www.vedomosti.ru/newspaper/article/243439/ne_nado_putina_ (accessed 25 March 2011).

Vezhin, S. (2011) ' "Edinaya Rossiya" Beret Novye Goroda', *Nezavisimaya gazeta*. Online. Available at http://www.ng.ru/politics/2011–02–28/1_edro.html (accessed 25 March 2011).

Vg-news (2009) 'V Khakhasii Edinaya Rossiya Vyvedet Na Miting 10 Tysyach Chelovek', Agenstvo Informatsionnykh Soobshchenii. Online. Available at http://www.vg-news. ru/news-v-khakasii-edinaya-rossiya%C2%BB-vyvedet-na-miting-10-tysyach-chelovek (accessed 25 March 2011).

Vinogradov, M. (2004) 'Etot Zapret – Izbytochen', *Izvestiya*. Online. Available at http://www. izvestia.ru/russia/article523404/ (accessed 23 March 2011).

Volkova, M. (2000) 'Odnim Kandidatom Men'she', *Nezavisimaya gazeta.* Online. Available at http://www.ng.ru/politics/2000–02–05/3_less.html (accessed 24 March 2011).

Von Beyme, C. (1985) *Political Parties in Western Democracies*, Trowbridge: Gower.

Vorob'ev, A. (2007) Chair of Central Executive Committee, United Russia, interview with author, 24 November, Moscow.

Vorob'ev, V. (2001) 'V Gosdume Proleteli Deputaty Ne Poderzhali Votum Nedoveriya Pravitel'stvu', *Trud.* Online. Available at http://dlib.eastview.com/sources/article. jsp?id=2262723 (accessed 2 June 2009).

Vorob'ev, V. (2003) 'Novosti: Sol'naya Partiya', *Rossiiskaya gazeta.* Online. Available at http:// dlib.eastview.com/sources/article.jsp?id=5743046 (2 June 2009).

VTsIOM (2008) 'Krisis Oppozitsiyu Ne Greet'. Online. Available at http://old.wciom.ru/arkhiv/ tematicheskii-arkhiv/item/single/11065.html?L%5B0%5 D=0%26c&cHash=e3a2e936ca (accessed 23 March 2011).

VTsIOM (2009) 'Poiski Hatsional'noi Idei Rossii Prodolzhayutsya'. Online. Available at http:// wciom.ru/index.php?id=266&uid=12780 (accessed 23 March 2011).

VTsIOM (2010a) 'Televidenie v Nashei Zhizni'. Online. Available at http://wciom.ru/index. php?id=268&uid=13518#« (accessed 23 March 2011).

VTsIOM (2010b) 'Doverie Politikam'. Online. Available at http://wciom.ru/index. php?id=169> (accessed 23 March 2011).

VTsIOM (2010c) 'Terroristy i Ikh Misheni: Reiting Opasenii Rossiyan'. Online. Available at http://wciom.ru/index.php?id=515&uid=13802 (accessed 23 March 2011).

VTsIOM (2011) 'Odobrenie Deyatel'nosti Obshchestvennykh Institutov'. Online. Available at http://wciom.ru/index.php?id=173 (accessed 23 March 2011).

Waldner, D. (1999) *State Building and Late Development*, Ithaca, NY: Cornell University.

Way, L. (2005) 'Authoritarian State Building and the Sources of Regime Competitiveness in the Fourth Wave', *World Politics*, 57: 231–61.

Way, L. (2006) 'Pigs, Wolves and the Evolution of Post-Soviet Competitive Authoritarianism, 1992–2005', Centre on Democracy, Development, and the Rule of Law, report number 62. Online. Available at http://cddrl.stanford.edu/publications/pigs_wolves_and_the_evolution_of_postsoviet_competitive_authoritarianism_19922005/ (accessed 26 March 2011).

Webb, P. (1996) 'Apartisanship and Anti-Party Sentiment in the United Kingdom: Correlates and Constraints', *European Journal of Political Research*, 29: 365–82.

Weber, M. (1946) 'Science as a Vocation', in H. Gerth and C. Wright Mills (eds), *From Max Weber: Essays in Sociology*, New York: Oxford University Press, pp. 129–56.

Weingast, B. (2003) 'A Post-Script to "the Political Foundations of Democracy in the Rule-of-Law" ', in A. Przeworski and J. Maravall (eds), *Democracy and the Rule of Law*, Cambridge: Cambridge University Press, pp. 109–13.

Weir, F. (2006) 'Echoes of Russia's Communist Past?'. *Christian Science Monitor*. Online. Available at http://www.csmonitor.com/2004/1005/p06s01-woeu.html (accessed 24 March 2011).

White, D. (2006) *The Russian Democratic Party Yabloko*, Aldershot: Ashgate.

White, J. (2006) 'Party Origins and Evolution in the United States', in R. Katz and W. Crotty (eds), *Handbook of Party Politics*, London: Sage, pp. 25–33.

White, S. (2000) *Russia's New Politics: The Management of a Post Communist Society*, Cambridge: Cambridge University Press.

Whitefield, S. (2001) 'Partisan Divisions in Post-Communist Russia', in A. Brown (ed.), *Contemporary Russian Politics: A Reader*, New York: Oxford University Press, pp. 235–43.

Whitefield, S. (2002) 'Political Cleavages and Post-Communist Politics', *Annual Review of Political Science*, 5 (1): 181–200.

Whitmore, B. (2008) 'Russia: The New and Improved Single-Party State', Radio Free Europe/ Radio Liberty (RFE/RL). Online. Available at http://www.rferl.org/content/article/1109574. html (accessed 224 March 2011).

Wiatr, J. (1995) 'The Dilemmas of Re-Organizing the Bureaucracy in Poland during the Democratic Transition', *Communist and Post-Communist Studies*, 28 (1): 153–60.

Widner, J. (1992) *The Rise of the Party State in Kenya*, Berkeley: University of California Press.

Wilson, A. (2005) *Virtual Politics: Faking Democracy in the Post-Soviet World*, New Haven, CT: Yale University Press.

Wyman, M., S. White, B. Miller and P. Heywood (1995) 'Public Opinion, Parties and Voters in the December 1993 Russian Elections', *Europe-Asia Studies*, 47 (4): 591–614.

Yanbukhtin, E. (2008) *Tekhnologii Uspeshnoi Izbiratel'noi Kampanii*, Moscow: Vershina.

Yashin, I. (2007) 'Politzavod "Edinoi Rossii" Obankrotilsya', *Novaya gazeta*, 12–14 October, pp. 7–8.

Zakaria, F. (2004) *The Future of Freedom: Illiberal Democracy at Home and Abroad*. New York: W. W. Norton.

Zakatnova, A. (2006) 'Proshchanie s Ogon'kom', *Rossiiskaya gazeta*. Online. Available at http://www.rg.ru/2006/10/21/prompartia.html (accessed 25 March 2011).

Zakatnova, A. (2007) 'Plan Pobedy', *Rossiiskaya gazeta*. Online. Available at http://www. rg.ru/2007/05/22/grizlov.html (accessed 25 March 2011).

Zolotov, A. (1999) 'Creating Reality', *The Moscow Times*. Online. Available at http://dlib.east-view.com/sources/article.jsp?id=236525 (accessed 5 January 2009).

Zubchenko, E. (2008) 'Platit' Ili Ne Platit'?', *Novye izvestiya*. Online. Available at http://www. newizv.ru/news/2008–07–10/93617/ (accessed 23 March 2011).

Index